Studies in Diplomacy and International Relations

The series was launched as *Studies in Diplomacy* in 1994 under the general editorship of G. R. Berridge. Its purpose is to encourage original scholarship on all aspects of the theory and practice of diplomacy. The new editors assumed their duties in 2003 with a mandate to maintain this focus while also publishing research which demonstrates the importance of diplomacy to contemporary international relations more broadly conceived.

Titles include:

G. R. Berridge (*editor*)
DIPLOMATIC CLASSICS
Selected Texts from Commynes to Vattel

G. R. Berridge, Maurice Keens-Soper and T. G. Otte
DIPLOMATIC THEORY FROM MACHIAVELLI TO KISSINGER

Herman J. Cohen
INTERVENING IN AFRICA
Superpower Peacemaking in a Troubled Continent

Andrew F. Cooper (*editor*)
NICHE DIPLOMACY
Middle Powers after the Cold War

Mai'a Davis Cross
THE EUROPEAN DIPLOMATIC CORPS
Diplomats and International Cooperation from Westphalia to Maastricht

David H. Dunn (*editor*)
DIPLOMACY AT THE HIGHEST LEVEL
The Evolution of International Summitry

Brian Hocking (*editor*)
FOREIGN MINISTRIES
Change and Adaptation

Brian Hocking and David Spence (*editors*)
FOREIGN MINISTRIES IN THE EUROPEAN UNION
Integrating Diplomats

Michael Hughes
DIPLOMACY BEFORE THE RUSSIAN REVOLUTION
Britain, Russia and the Old Diplomacy, 1894–1917

Gaynor Johnson
THE BERLIN EMBASSY OF LORD D'ABERNON, 1920–1926

Christer Jönsson and Martin Hall
ESSENCE OF DIPLOMACY

Donna Lee
MIDDLE POWERS AND COMMERCIAL DIPLOMACY
British Influence at the Kennedy Trade Round

Donna Lee, Ian Taylor and Paul D. Williams (*editors*)
THE NEW MULTILATERALISM IN SOUTH AFRICAN DIPLOMACY

Mario Liverani
INTERNATIONAL RELATIONS IN THE ANCIENT NEAR EAST, 1600–1100 BC

Jan Melissen (*editor*)
INNOVATION IN DIPLOMATIC PRACTICE

THE NEW PUBLIC DIPLOMACY
Soft Power in International Relations

Peter Neville
APPEASING HITLER
The Diplomacy of Sir Nevile Henderson, 1937–39

M. J. Peterson
RECOGNITION OF GOVERNMENTS
Legal Doctrine and State Practice, 1815–1995

Gary D. Rawnsley
RADIO DIPLOMACY AND PROPAGANDA
The BBC and VOA in International Politics, 1956–64

TAIWAN'S INFORMAL DIPLOMACY AND PROPAGANDA

Karl W. Schweizer and Paul Sharp (*editors*)
THE INTERNATIONAL THOUGHT OF HERBERT BUTTERFIELD

Ronald A. Walker
MULTILATERAL CONFERENCES
Purposeful International Negotiation

A. Nuri Yurdusev (*editor*)
OTTOMAN DIPLOMACY
Conventional or Unconventional?

Studies in Diplomacy and International Relations
Series Standing Order ISBN 0–333–71495–4
(*outside North America only*)

You can receive future titles in this series as they are published by placing a standing order. Please contact your bookseller or, in case of difficulty, write to us at the address below with your name and address, the title of the series and the ISBN quoted above.

Customer Services Department, Macmillan Distribution Ltd, Houndmills, Basingstoke, Hampshire RG21 6XS, England

---

The International Thought of Herbert Butterfield

Edited by

Karl W. Schweizer

Professor
Department of Humanities
New Jersey Institute of Technology, USA

and

Paul Sharp

Professor of Political Science
University of Minnesota Duluth, USA

palgrave
macmillan

First published 2007 by
PALGRAVE MACMILLAN
Houndmills, Basingstoke, Hampshire RG21 6XS and
175 Fifth Avenue, New York, N.Y. 10010
Companies and representatives throughout the world

PALGRAVE MACMILLAN is the global academic imprint of the Palgrave Macmillan division of St. Martin's Press, LLC and of Palgrave Macmillan Ltd. Macmillan® is a registered trademark in the United States, United Kingdom and other countries. Palgrave is a registered trademark in the European Union and other countries.

ISBN-13: 978–0–230–00166–4 hardback
ISBN-10: 0–230–00166–1 hardback

This book is printed on paper suitable for recycling and made from fully managed and sustained forest sources.

A catalogue record for this book is available from the British Library.

Library of Congress Cataloging-in-Publication Data
The international thought of Herbert Butterfield / edited by
Karl W. Schweizer and Paul Sharp.
p. cm.
Includes bibliographical references and index.
ISBN-13: 978–0–230–00166–4 (cloth)
ISBN-10: 0–230–00166–1 (cloth)
1. Butterfield, Herbert, Sir, 1900– 2. International relations—Philosophy. I. Schweizer, Karl W. II. Sharp, Paul, 1953–
D15.B86I I68 2007
327.1092—dc22 2006051589

10 9 8 7 6 5 4 3 2 1
16 15 14 13 12 11 10 09 08 07

Printed and bound in Great Britain by
Antony Rowe Ltd, Chippenham and Eastbourne

Contents

Foreword

For those who count themselves disciples of Sir Herbert Butterfield, this volume is long overdue for readers on both sides of the Atlantic. For some it will reawaken fond memories of meetings with Sir Herbert and lively correspondences that ended only with his passing. One such memory that remains in my mind is of my participation, albeit brief, in the British Committee On the Theory of International Politics, with Sir Herbert chairing. He did so with both firmness and the deepest form of understanding of what others were trying to develop. The results for me were abiding friendships that survived his death with people who came to lecture in my classes at the University of Virginia and to teach at the University of Chicago. Something happened at Peterhouse when Butterfield was both Master of the College and Chair of the Committee that I do not recall happening at any other British or American university. Our meetings would begin on Friday at noon in a relaxed manner with Sir Herbert posing a succession of 'Butterfield questions' to which he invited us to respond. Some of these would become unifying themes for the discussion which would continue until noon the following Monday. None of us could explain how it all worked, but you will get a sense of the approach as you read the following pages and feel yourself tested by the questions and arguments of a small, wiry British professor, just as we did all those years ago.

The pieces are organised around four main topics: European Diplomatic History; Faith and Ethics in International Politics; International Theory; and Diplomacy. Sir Herbert attacks diplomatic history but remains convinced that, done properly, it provides the best understanding of what remains, at heart, a very human business. His writings on faith and ethics provided those of us interested in morality and foreign policy sub-texts for another ten or fifteen years. His contributions to international theory demonstrate the extent to which the divide between classical and behavioural approaches was always overdrawn. Finally, in his section on diplomacy he rejects the common cynicism about the profession and concludes that a good diplomat is very likely to be a good person.

Throughout the volume, we can see Sir Herbert's quest for unifying themes in our understandings of international affairs, what are they like, how might they be, and what should we make of them. These are themes that remain as relevant today as they were three-quarters of a century ago. Of how many scholars and publicists active today is this likely to be said?

Kenneth W. Thompson
University of Virginia

Acknowledgements

First and foremost, the editors with to thank Mr Peter Butterfield, literary executor of the Butterfield Papers for allowing us to consult his father's manuscripts and granting permission to print the previously unpublished essays in this volume. We also wish to pay tribute to the memory of his mother, the late Lady Pamela Butterfield, who had enthusiastically supported the project from the beginning. Among the scholars who shared with us their critical insights, and in some cases, personal knowledge of Sir Herbert Butterfield we are grateful to Adam Watson, C. T. McIntire, G. R.Berridge, Michael Bentley, Richard Langhorne, Brunello Vigezzi, Keith Sewell and Phillip Towle, Roger Lovett and Brendan Sims at Cambridge. Special thanks are due to Professor Kenneth Thompson, long time friend of Sir Herbert, for agreeing to write the Foreword. We are likewise indebted to the following scholarly institutions: the Centre of International Studies, Cambridge University; the Cambridge University Library; the Institute of Historical Research, London; Princeton University Library; the University of Minnesota Duluth Library and the Rutgers University Library, Newark, New Jersey for complying with our innumerable requests. Research was made possible by the generosity of the New Jersey Institute of Technology and the University of Minnesota (Duluth) both of whom provided a semester sabbatical leave, and Peterhouse College, Cambridge, who elected one of the editors a Senior Research Associate during that period of leave. We also wish to express our gratitude to Dr Fadi Deek, Dean of the College of Science and Liberal Arts, at NJIT, for generously helping to defray typing costs, to Dr Robert Lynch, Chair, Department of Humanities at the same institution for ever willing help and cooperation; and to Dean Linda Krug at the College of Liberal Arts at the University of Minnesota Duluth for Assistance and encouragement. We further thank Alison Howson, acquisition editor and Gemma D Arcy Hughes, editorial assistant at our publisher, Palgrave McMillian, for their admirable patience and ready assistance at different stages in the preparation of this project. We also owe a tremendous debt of gratitude to Don Spidell, III for his invaluable technical assistance in formatting the final version of the manuscript. Last but not least we wish to express our deep-felt gratitude to Pamela Spidell for her exceptional computer and copy-editing skills; from first to last she has proven a model of resourcefulness, perception, perseverance, and good humour.

The editors and publishers have made every effort to contact copyright-holders. If any have inadvertently been overlooked, the publishers will be pleased to make the appropriate arrangements at the first opportunity.

Introduction

Sir Herbert Butterfield (1900–79) was a leading British historian who spent his career at Cambridge University. By training, Butterfield was a diplomatic historian, but his interests led him towards 18th century British political history, historiography, the history of science and international political theory. He wrote and published widely, but is now chiefly remembered for his attack on progressive or Whiggish history and, more narrowly, for his attempts to make sense of the international affairs of the Cold War from a Christian perspective through a series of public broadcasts and lectures.[1] Our purpose in this volume is to bring together Butterfield's best work, both published and unpublished, on international thought. We use this term broadly to cover his diplomatic history narratives and essays on the great international issues of his time, as well as his international thought in the narrower more philosophical or theoretical sense. 'All well and good ...' our commissioning editor responded when we brought her our proposal, but, she asked, '... why publish a Butterfield collection now, and what are you going to include?'

Neither question is easy to answer for, in Butterfield, we have a diplomatic historian who actually wrote very little diplomatic history, a historiographer whose forays beyond his lifetime' fascination with Acton and the German School revealed little more than the embryonic state of the history of historiography, a Christian moralist who thought that the USSR might have more than the West to offer the latter's former colonies, and a theorist who seemed to understand 20th century international conduct in terms of a fall from 18th century diplomatic grace precipitated by the events of 1914.[2] As his biographer notes, Butterfield tended to identify two aspects – neither more nor less – to any intellectual problem and move back and forth between them over the course of his life.[3] Thus, he was an advocate for technical history capable of passing the most stringent evidentiary tests, yet he maintained that a historical imagination and, indeed, a talent for fiction, were more useful than mere facts for understanding a historical episode and getting at the truth.[4] He was an advocate of diplomatic history – the careful accounting of the deliberations and actions of the powerful and those who served them – yet maintained that such a focus, neglecting as it did the social and economic contexts which shaped understandings and choices, missed a vital part of the story.[5] Most strikingly, in these terms, he regarded himself as a near-pacifist and was a firm (albeit *soto voce*) advocate of nuclear unilateralism, but seemed fascinated by the maneuvers and calculations of high politics and had little sympathy for those who depreciated what he saw as the inescapable presence of power in international affairs.[6]

It was not the existence of such tensions with which subsequent critics sometimes took issue, but with Butterfield's apparent lack of interest in resolving them. Instead, he appears to have taken, what his biographer calls, a dissenting approach, questioning and undermining what he took to be the prevailing consensus on questions that attracted his interest at any given time. Thus, for example, when diplomatic history enjoyed a complacent ascendancy he led the attack on it, but when it was close to dead and buried he sought to remind others of its virtues. Pushed to the point on difficult questions, Butterfield advocated at times a kind of Augustinian quietism and refusal to commit which offered certain comfort only to people of faith. It was easier, his critics suspected, to trust in Providence and to be prepared, in principle, to embrace catastrophe, when Providence had been so good as to make you the Master of Peterhouse and the prospect of actual catastrophes to seem remote.

Paradoxically (although he would not have regarded it as a paradox), it is Butterfield's failures and difficulties in narrow scholarly terms, together with his ability to communicate them clearly and openly, which accounts for much of the continued interest in him. His encounters and re-encounters with basic questions of life across the whole sphere of human endeavour, coming back again and again to the same questions, rather than building beyond them, accords with the way most of us live our lives. It might be claimed, therefore, that Butterfield's real *métier* was as a publicist, the equivalent of an opinion writer for the policy journals or quality papers today, only working in the older media of radio broadcasts and public lectures. If so, one can only marvel at his ability to bring people up to the levels at which complex and difficult issues need to be discussed, and contrast this favorably with the present tendency to boil things down to the point at which, it is believed, that the attention of ordinary people may be momentarily retained.

Of course, Butterfield was far more than an effective publicist (even when he was engaged in stimulating reflection and debate among the public). As is attested to by his distinguished career, by his continuing use as a marker in debates about progress in history and historiography, and by the relatively small band of professional historians and even smaller (though growing) cardre of international relations theorists who have retained his orientations to their respective fields, Butterfield was a first-rate scholar. To this constant standing, however, has been added a pronounced revival of interest in recent years manifested in the intellectual biography referred to above, a treatment of Butterfield's historical method, and several scholarly articles.[7] What all of these reflect is a discovery by scholars in several fields that Butterfield had all along been formulating problems in terms which for years had been unfashionable, but in which they have newly become interested. As far as the study of international relations is concerned, these may be summarised in the following themes:

- an interest in narrative accounts which seek to understand the terms in which people, as historical subjects, see themselves acting;
- an emphasis on people as individual personalities rather than as individual identities and egos, bearers of social structures, or constituent parts of collective social agencies;
- an acknowledgment of the spiritual dimension to human life as a sociological fact and, more controversially, as a metaphysical and transcendent reality;
- a privileging of the attempt to incorporate certain virtues – notably humility and charity – in individual human conduct over claims about the right ordering of societies in terms of justice and attempts to establish those right orders;
- an advocacy of the virtues of diplomacy and diplomatic conduct understood as expressing all of the above in practice.

The ways of understanding and acting implied by these themes have been neglected in the contemporary study of international relations, and it is the position of the editors that their revival is long overdue.

The case for why anyone should look at Butterfield is, of course, best made by the writings themselves. This, however, brings us to our editor's second question, '... what are we going to include?' As people familiar with Butterfield's work are well aware, he was a prolific writer. Given his ability to develop common themes in his treatment of widely disparate areas of human conduct, restricting this collection to pieces specifically about international affairs cut out a great deal of what is relevant to his international thought. Thus, the problem which presented itself to the editors was not one of what to include, but what to leave out. It is a safe bet that the first substantive criticism reviewers will make of this volume is that they cannot believe that the editors neglected this or that piece which they regard as important. We accept the criticism in advance, for we have agonised profoundly ourselves over some of the things that have been left out.

A second, less important, but in its own way, more difficult problem to solve is Butterfield's tendency to repeat themes, arguments and illustrative examples. Both the prolific character and repetitive element in his work arose in part because, as his biographer notes, he thought by writing and because he over-committed himself to public presentations and popular collections of what he had to say.[8] When one sees the number of drafts of essays on, for example, Western Civilisation, Just War, or the Balance of Power which appear in his files, it is small wonder that some of his great projects, notably the planned biography of Fox and the oft-promised general history of Europe, were never brought to completion. Like writing, however, repetition was part of Butterfield's thought process, not least because he addresses problems which admit of no definitive or final answer, but which bother us, will not go away, and therefore must be confronted.

The problem of repetition manifests itself in Butterfield's unpublished writings and especially in earlier drafts of pieces which were subsequently published. Usually, we were able to agree on the strongest iteration of a particular argument and usually this was the published version where such existed. On occasions, however, we were faced by similar drafts making different points of broadly equal value. Constraints of space entail that we have not presented both or multiple drafts of the same idea so that the reader can tease out the differences and developments in Butterfield's thinking which these may reflect. Rather, we have made judgments about which is the stronger version and, where necessary, have inserted our own commentary on, or excerpts from, the unused draft. We are still not satisfied that this approach was completely effective but adopted it for want of a better alternative.

More generally, we answer the question of what to include by organising the collection around a very rough characterisation of the thematic development of Butterfield's international thought. The volume begins with samples of his diplomatic history, the sub-field in which he originally made his name, and his commentaries on diplomatic history, of which considerably more exists.

The second section deals with Butterfield's commentary on ethical aspects of international relations that, of necessity for him, involve questions of religious commitment in general and Christian faith in particular. How should governments behave in international relations and, given how they actually do behave, how should Christians respond? Here, he attempted to clarify the providential and historical dimensions of human existence and to create a renewed awareness among statesmen of the practical wisdom Christian precepts can offer. This section also deals with what were taken to be the big issues of international relations in the 1950s and 1960s. What should be done about the Communists? What should be done about those seeking to maintain their neutrality in the Cold War? And, most important of all, what should be done about our own governments in the West, their self-righteousness and hubris? The examples are dated but, as we hope to show, they illustrate important problems, the character of which remains fundamentally the same. They also demonstrate how, if Butterfield is to be regarded as a conservative, his conservatism was of a strange and potentially very radical sort. The acclaim by conservatives he enjoyed in some parts of the United States in the 1960s, for example, would not be repeated in the official America of George W. Bush.

The third section focuses on Butterfield's contributions to, and views of, the work of the British Committee on the Theory of International Politics. Again there are surprises, for the Butterfield who declares sovereign individual personalities to be the proper subject of narratives written so that we can come to understand people in their own terms,[9] is revealed as open to the study of international relations as a science. Personalities may be sovereign, but Christian charity insists that we give an account of the circumstances and structures, as he called them, which influence people's choices. Butterfield encouraged 'scientific' inquiry directed at the elaboration of such constraints and saw the British Committee as making a contribution, in principle at least, to the maintenance of a stable international order capable of undergoing peaceful change.[10]

Finally, we bring the selection full circle by presenting a selection of Butterfield's writings on diplomacy rather narrowly defined. It is here, perhaps, that we take our greatest liberty as editors for, in ordering the selections thus, we advance our own thesis about Butterfield, namely that he saw the conduct of good diplomacy, as he understood it, as a model for good conduct in general. The world would be a better place if more people thought and acted like good diplomats. Back then to diplomatic history in the curriculum, not just to learn about the international relations of the past but, in so doing, to obtain proper intellectual and moral training.[11]

Notes

1. *The Whig Interpretation of History*, Cambridge, 1931. A select bibliography of Butterfield's works may be found on p.xxx. A complete bibliography of Butterfield's published and unpublished works may be found in C.T. McIntire, *Herbert Butterfield: Historian as Dissenter*, Yale, 2004.
2. *Peace Tactics of Napoleon, 1806–1808*, Cambridge, 1929 is Butterfield's only full-length study of diplomatic history, and even this has attracted some criticism for its use of secondary sources. See *MacIntire* pp. 386–401 for a critical, but appreciative, commentary on Butterfield's work on the history of historiography, especially as this is found in Butterfield's *Origins of History* edited by Adam Watson, London, 1981. For comments on Communism, societies' ills and how to order former colonies see 'Morality and International Order' in Brian Porter (ed.), *The Aberystwyth Papers: International Politics 1919–1969*, London, 1969 and 'Human Nature and the Dominion of Fear' in *International Conflict in the Twentieth Century*, New York, pp.79–98. For perspectives on Butterfield's presentation of 18th century diplomacy and international relations see Tim Dunne, *Inventing International Society: A History of the English School*, Basingstoke, 1998, pp.71–88 and reviews of *Christianity, Diplomacy and War*, London, 1953 by Martin Wight, *Observer*, 16 August 1953 and A.J.P. Taylor, *Manchester Guardian Weekly*, 13 August 1953, both reviews in File 99, Butterfield Collection, Cambridge University Library, (hereafter B.C.).

3. *McIntire*, pp.193–5. cf J. Derry, 'Herbert Butterfield and the Interpretation of History,' *English Historical Review*, cxxi (2006), pp. 258–68.
4. *History and Human Relations*, London, 1951 pp.69, 74 and especially 'Moral Judgments in History' pp.101–30, *Man On His Past: The History of the Study of Historical Scholarship*, Cambridge 1955 throughout, 'The Establishment of a Christian Historiography' in *The Origins of History*, p.165, and *The Historical Novel: An Essay, Cambridge*, 1924, throughout.
5. See for example, 'In Defence of Diplomatic History,' File 426 B.C. and 'Marxist History' in *History and Human Relations*, pp.66–100.
6. On nuclear weapons see, for example, 'Human Nature and the Dominion of Fear' in *International Conflict in the Twentieth Century* and Butterfield's correspondence with the Reverend A.J. Muste, 24 November 1960 and Mrs. Aitken, 1 May 1964, both in File 425 BC. For the inescapability of power politics see 'Diplomacy' in Ragnhild Hatton and M.S. Anderson, *Studies in Diplomatic History: Essays in Honour of David Bayne Horn*, London, 1970, pp.357–73. K. Sewell, 'The Inescapable Predicament: Sir Herbert Butterfield's Reflections on the Human Dilemma,' *Fides et Historia*, vol. 16, (autumn 1983) 7–17. For Butterfield's attraction to 'power' politics see his own comments on Martin Wight's view of him as a 'wangler' in *The First Martin Wight Memorial Lecture: Raison D'Etat: the Relations Between Morality and Government*, File 111BC, 1975 and published version of same, Brighton, 1975.
7. For historians see the work of Michael Bentley, especially 'Butterfield at the Millennium: the "Sir Herbert Butterfield Lecture," 1999,' *Storia della Storiografia*, 38 (2000), 17–32 and Keith C. Sewell, *Providence and Method: Herbert Butterfield and the Interpretation of History*, Sioux City 2002. For international relations theorists see Roger Epp, 'The "Augustinian Moment in International Politics": Niebuhr, Butterfield, Wight, and the Reclaiming of a Tradition,' *International Research Papers*, no. 10, Aberystwyth: Department of International Politics, University College of Wales, 1991 and Ian Hall, 'History, Christianity and diplomacy: Sir Herbert Butterfield and International Relations,' *Review of International Studies*, 28, 2002, pp. 719–36.
8. *McIntire*, pp. 403–17.
9. See for example, *Christianity and History*, New York, 1950, pp.66.
10. See, for example, Butterfield's contribution to the balance of power discussion at the Bellagio conference in 1967 on international relations theory, File 109 BC.
11. *History and Human Relations*, p.91; *Man on His Past*, p.xiv.

Part I
Diplomatic History

1
Introduction

The essays in this section are representative samples of Butterfield's lifelong preoccupation with and exploration of the moral problems inherent in history, politics and statecraft viewed from a Christian perspective – themes that have been an integral part of the political life of Western Civilisation since its advent. In his quest for, what one writer has called the 'nature of wise statesmanship,'[1] Butterfield characteristically relied upon his historical training rather than abstract theorising, for insights, precepts and guidance. Particularly influential here was the field of diplomatic history, an area of scholarly inquiry resting on imposing intellectual foundations. Building upon the Rankean tradition – inherited from his teacher Harold Temperley[2]– he believed that diplomatic history was not only an ideal preparation for aspiring historians 'because of the precision work that it required,'[3] but was a tapestry of knowledge, discernment and perspective that broadened the mind, enhanced human understanding and refined the intellect: a process essential for historical exposition of a higher order as well as more prudent political decisions.[4] Good diplomatic history and good diplomacy are based on a shared set of skills and virtues and ultimately, all history has a diplomatic dimension to it. Indeed, as Butterfield once observed '... we shall find at the heart of everything a kernel of difficulty which is essentially a problem of diplomacy as such'[5]. The discipline, according to Butterfield, was also a vital point of departure for diplomatic theory, trenchantly expressed in such classic works as Callières, *De la Manière de négocier* (1716), Martens' *Guide Diplomatique* (1832) and Nicolson's *The Evolution of Diplomatic Method* (1954). Throughout all of these (*inter alia*) the history of diplomacy reveals maxims whose enduring validity was unconnected with any particular time or place: in other words, was fundamental to the unchanging nature of the human personality and condition, this connecting the past with the complexities of the present.

In the first paper which follows, 'In Defense of Diplomatic History', initially delivered to the Cambridge University History Club (November 1963), Butterfield reviews the subject's historical development, the reasons for its gradual eclipse, but most importantly he explores its themes, modalities and enduring relevance for scholars and policy makers alike. Besides teaching the techniques of research, diplomatic history, it is argued, offers insights both into a state's foreign policy and its domestic affairs; at some level also into its culture and self-perception. As such, diplomatic history provides an entrée into the world of ideas – those conscious purposes and reflections which lend sense, structure and continuity to political events and guide humanity in the light of past experience. It yields the important benefit of counteracting the type of approach which takes events out of their human context: reduces the wealth of human practice to disembodied abstractions and formulas. Such schematisation risks eliminating – or actually seeks to eliminate – the intractable elements in decision making, reducing humans to automatons, their actions, feelings and personality viewed in a vacuum. As he emphasises, 'I want to be sure that the world does not lose sight of the other side of the truth: the role of human rationality and will – the importance of the decisions men actually make.'

The essay, in addition, reminds readers that diplomacy, at all times, operates against a background of tensions emanating from power and interest, and therefore, entails more than simply reasonable dialogue: a case of minds 'engaged in argumentation.' It also has to address the vexing problem of human intractability. And because intractability is involved the spectre of power is involved. Yet without recourse to diplomacy, 'there would be nothing left except the appeal to force. Even amidst negotiation, power is ever grimly present, though attenuated, diplomacy making it . . . change shape and operate through a different channel . . . in a more refined state and at a different temperature.' In this sense, diplomacy represents the civilizing influence of reason on what would otherwise be ceaseless turmoil: a Hobbesian jungle – *bellum omnium contra omnes*.[6] Here diplomatic history plays a clarifying role: encouraging the analysis of crises and issues on elevated levels of perception that identify and dissect what Butterfield elsewhere calls the 'tragic element' in modern conflict: an endemic human predicament inherent in the very 'dialectic of events,' regardless of unusual wickedness or criminality.[7]

The importance of individual personalities, their interplay and friction, for international politics – as opposed to abstract forces – is illustrated in the second paper, originally the opening chapter of Butterfield's major foray into diplomatic history, *The Peace Tactics of Napoleon* (1929), supervised by Harold Temperley. Written in flowery prose, based on English, French and Austrian archival material, this work was a meticulous exploration of the diplomatic manceuvers accompanying the military campaigns of the 4th coalition against Napoleon culminating in the meeting between Czar Alexander I and the Emperor at Tilsit in early July 1807. Especially interesting here is Butterfield's dramatic evocation of the human dynamics – the intricate pattern of calculation, opportunism, caprice and pique shaping Napoleon's diplomacy following his defeat of the Prussians at Jena. He concludes, an early intimation of his later position,[8] that diplomacy, creatively co-exists with war, that correspondingly even peace – or offers of peace – may be a '*ruse de guerre*' in any age of history'[9] and that the best diplomatic history is the kind which not merely codifies events – mechanically dovetails documents – but graphically resurrects them in their full narrative context, as living realities with proper emphasis on the hidden undercurrents and clash of personalities involved. By implication the paper also adumbrates the point, made in section four of this volume, that diplomacy is a truly creative enterprise in the hands of consummate statesmen – a combination of fertile expedients calling for the ultimate in mental flexibility, tactical perception, judgment, moral sensitivity and if necessary, cajolery – anything short of outright war.[10]

The third paper (book IV, chapter III from the same volume) vividly describes the implication of ministers at odds with their own government on critical policy issues and the importance of 'diplomatic culture' if not technique – in this case, allied diplomats restraining their personal feelings against Napoleonic tyranny, silently concocting counter strategies with subtlety and precision, while waiting patiently for Napoleon's excesses to precipitate his downfall. Altogether the paper effectively captures the human dimension at the heart of international affairs, its impact on outcomes, and the way in which good diplomats have a fine sense of timing, patience and restraint concerning the possibilities for effective action. Moreover Butterfield shows, literally, how diplomatic history, written with freshness, vigour and sympathetic imagination, can recover the immediacy of past events without sacrificing the integrity of history as a scholarly discipline requiring rigorous and detailed research. The turbulent course of world affairs in the 1930s and 1940s inspired Butterfield to incisive reflections on the nature of international politics and to apply his prescriptions for creative politics and wise statecraft. He saw the First World War as the great divide: a cataclysmic event with disastrous repercussions for Britain and Europe at large: what fractured the continuity of European civilisation and triggered chain reactions culminating in further calamities to follow.

Studying the crisis of 1914 from his characteristically nuanced historical stance, Butterfield confronted the question whether war had been inevitable or whether 'through the creative statesmanship and vision of a few people it could have been averted.'[11] The issue also intensified his endemic preoccupation with historiography, the pre-history of the war illustrating the fertile interaction between settled assumptions, discrepant evidence and new narrative structures: scholarship moving to higher levels of analysis. His critical research on this topic first appeared in a paper delivered to 'The British Committee on the Theory of International Politics' in 1960 – a paper entitled Crowe's Memorandum of 1 January 1907, which represents the fourth selection from his writings on diplomatic history printed below.[12] Its focus is the debate between pro-German and anti-German biases in British foreign policy before 1914, and how this debate was conditioned by historical perspectives at once too confined, too rigidly fixed, too rooted in official orthodoxy. Here Butterfield strongly emphasises a point also made 'In Defense of Diplomatic History' – namely, the dangers for decision makers of exclusive preoccupation with current history and assumptions, as these foster parochialism, distort perspectives and prompt statesmen 'to lock themselves more securely in the deceptions and illusions of their age.'[13] This he concretely illustrates by comparing Crowe's Germano-phobic memorandum with the riposte written by Lord Sanderson the following month, which advocated sensitivity, tact and the need for a more balanced historical overview that recognised enmities, competition and provocations on all sides, not just Germany – recognised also the importance of viewing interstate relations from a higher altitude: 'seeing ourselves as others see us'. Butterfield, moreover, praised Sanderson for perceiving the true tendencies of state behaviour – all states – across a broad historical spectrum: the natural proclivity to seek security and to expand, given the opportunity. In addition, a new power, as Germany then was, tends to be nervous, impatient, susceptible to overreaction in their diplomacy – a fact that requires imaginative sympathy; something Sanderson clearly understood. Crowe, by contrast, saw only hostile purpose and intransigence, limitations of insight which inevitably compounded an already tense crisis situation. Good diplomacy, Butterfield reminds us, depends upon extensive historical knowledge for such breadth of knowledge enables the policy maker to transcend facile partisanship – to understand all sides locked in conflict and appreciate that more often than not, collisions are the product of what he calls 'the tragic element' referred to above – a universal human predicament, arising from situations themselves, a trackless wilderness, independent of consciously malevolent purpose.[14]

The final paper in this section, 'Sir Edward Grey in July 1914' originated as a lecture given during a tour of German universities in 1964 and subsequently expanded.[15] Based largely on the official British volumes of state papers edited by H. Temperley and G. P. Gooch, the essay, while acknowledging Germany as a threat to peace in 1914, makes Russia equally, if not more culpable, given her expansionism and belligerence long before the outbreak of war. Essentially he condemns Grey for adopting a 'moralistic' as opposed to a more 'scientific' attitude – one which recognised, in line with prudent statesmanship, the need for flexibility and mental agility – for surveying the European disposition of forces, calmly, comprehensively and most important, from a broad historical perspective. Grey, according to Butterfield, lacked any historical sense of the relationship between Germany and Russia and was too focused on Western Europe – especially the Anglo French entente. This restricted his horizons, made him blind to the stabilising potential of Russo-German fears,[16] and helped foster the polarisation of alliance groupings that foreclosed 11th hour diplomatic initiatives and so escalated the conflict spiral leading to actual war. Ultimately the paper on Edward Grey, and to varying degrees, the other papers in this section illustrate one of Butterfield's pre-eminent maxims: both in the realm of diplomacy and historical inquiry, imagination – sheer insight – is, in the final analysis, the greatest asset of all.

Notes

1. A. R. Coll, *The Wisdom of Statecraft: Sir Herbert Butterfield and the Philosophy of International Politics* (Durham, NC 1985) ch. I, pp. 4–6.
2. Cf John Fair, *Harold Temperley: A Scholar and Romantic in the Public Realm* (Newark, DE 1992) ch. VIII; C. T. McIntire, *Herbert Butterfield: Historian as Dissenter* (New Haven, 2004) pp. 22–40; 170–2; 254–5.
3. H. Butterfield, introduction to H. Temperley, *The Foreign Policy of Canning* (London, 1968), XVI.
4. 'The Historian and his evidence' BC File 128; H. Butterfield, *The Study of Modern History* (Cambridge, 1944) pp. 7–18.
5. H. Butterfield, *History and Human Relations* (London, 1951) p. 26.
6. The preceding paragraph is partly based on: K. W. Schweizer, 'A Case for Diplomatic History: Reflections from the Butterfield Archive', *Archives*, XXX, no. 113 (Oct. 2005) pp. 66–74.
7. H. Butterfield, *History and Human Relations* (London, 1951), pp. 17–29. idem, 'Morality and Process in international affairs' Butterfield Papers (BC) File 31.
8. H. Butterfield, 'Diplomacy' in: R. Hatton and M. S. Anderson eds *Studies in Diplomatic History: essays in memory of David Bayne Horn* (London, 1970) ch. 19. idem. 'The Development of Diplomacy' Butterfield Papers (BC) File 137.
9. H. Butterfield, *The Peace Tactics of Napoleon* (preface to Octagon reprint, 1972).
10. For a fuller elaboration of this see: Paul Sharp, 'Herbert Butterfield, the English School and the civilizing virtues of diplomacy,' *International Affairs*, 79, 4 (2003) pp. 855 – 78.
11. Coll, op.cit., p. 107.
12. It was presented in July of that year. (Vigezzi, op.cit., p. 328).
13. H. Butterfield, *The Study of Modern History* (London, 1944) p. 29.
14. H. Butterfield, *History and Human Relations* pp. 17 – 29.
15. McIntire, op.cit., p. 356.
16. A point he had already made in a paper delivered at Dublin on 14 December 1948, observing 'while Russia and Germany have been alternative menaces... we failed to widen our vision – failed ever to think of more than one of these possible menaces at the same time, failed to envisage two possible enemies and dangers at once, or even to see how they might be used as a mutual check – used even to cancel one another out.' H. Butterfield, 'Official History: Its Pitfalls and its Criteria.' *Historical Studies: An Irish Quarterly Review of Letters, Philosophy and Science*, 38 (1949) p. 140.

2
In Defence of Diplomatic History[1]

I get the impression that people are sometimes sad because the world has taken no notice of their ideas or their preaching; but I seem to be of a different temperament. I begin to be terrified when anybody starts to take notice of any suggestion of mine; and when I see a tendency to follow up one of my theories, I feel certain that I am going to be exposed – that something must be wrong. In one sense my apprehensions have proved to be well-founded. Some of the intellectual tendencies which prevailed when I was young and which I attacked, are now in such a bad way that I want to rescue them. I once opened a campaign against the Whig interpretation of history,[2] but this was taken up without some of the safeguards that I tried to establish and, like everything else, once it was taken up it was carried too far. In my young days, I was a severe critic of diplomatic historians, and it was very much on my insistence that the Modern European History paper in our Tripos was redefined to comprise the development of our civilisation – the relations between religious, political, intellectual, social and economic movements – instead of dealing very largely with the external relations of states. I am unable to convince myself that I was wrong in all this; but I begin to wonder whether I was as right as I thought. I remember the days when I was criticised and ridiculed because, instead of just taking the line that men create their own history by the decisions they make, I drew attention to the kind of history making that goes on over their heads, twisting their purposes and bringing their actions to results they never intended. But even this has gone sour on me: for the people who ought to have been nearest to me have pressed it so far as to knock the sense out of history and deny the role of human reason in the narrative. Now I want to be sure that the world does not lose sight of the other side of the truth – the role of human rationality and will – the importance of the decisions men actually made.

When a more scientific kind of historical study began to assume importance in the 19th century, diplomatic history played a great part in the story and came very much to the forefront. This was partly because the whole movement was closely associated with the opening of government archives, and for a considerable time it was the archives of the Foreign Office that held the field. They were convenient to work with – you could get in a nicely bound volume the correspondence between the British Foreign Secretary and the ambassador at a foreign capital (Vienna shall we say) from the first of January to a given date in a certain year. It was in the realm of foreign policy that governmental work had been particularly secret, so that historians were anxious to have access to the documents – here were lots of sensations, lots of historical revisions to be made. And the documents of Foreign Offices came into such general service that they were used even for the internal history of the various countries. The historians learned what foreign diplomats in London had reported about English politicians, English party controversies, and English economic life.

The most famous of all types of diplomatic document – in fact, I think the most famous single class of documents ever to appear in the modern history field – was a great group that emanated from Venice, particularly in the 16th and 17th century – what we call the Venetian relazioni,[3] the Venetian relation, a type that does not possess any very close parallel in more recent times. They were not ordinary diplomatic dispatches but full descriptions that ambassadors had to write about the countries in which they had completed their residence of two or three years. Though they were supposed to be very confidential, some of them leaked out early from the Venetian archives and were treasured as very distinguished documents just before and just after the year 1600. At the beginning of the 19th century, they were rediscovered and a little later the great German historian Ranke saw their special utility, hunted them in the archives and gained from them the first hint of the possibility of writing history entirely from manuscript sources. Venice was a commercial oligarchy and the Venetian diplomats had sufficient business sense to be interested in the economic life of the country in which they had resided. The economic history of Tudor times would be written from the descriptions produced by Venetian ambassadors; and these relazioni were literary documents – they were actually declaimed by the ambassador to the elder statesmen of Venice after he returned from his mission in a foreign country. He knew that his prestige would partly depend on the skill that he showed on such occasions. The high point in his account would be the full dress portrait he would give to the monarch and statesmen in the country in which he had resided. It was important for historians to have these vivid word pictures of men like Philip II of Spain produced by Venetian ambassadors who had seen him at close quarters.

But, apart from the technical importance of the diplomatic documents in the newly opened archives, there was a deep, organic reason why diplomatic history – the study of foreign relations – acquired a presidential position. Academic history – the serious planning of history as an academic study – emerged in the University of Göttingen, the most famous of the German universities in the later decades of the 18th century. Because Göttingen was in Hanover, its connections with England were close and it was influenced by the more liberal English outlook. It was interested in politics and it became a famous school of statesmanship, a place to which the sons of the nobility from all over Germany would come for a training particularly adapted to public life and government service. The study of states was an important part of this educational programme; and the Professors in Göttingen, as they moulded their scheme of European history, concentrated on state action and the history of the European states-system. The conduct of relations with other states – including the waging of wars – was seen as the highest kind of state action, the most momentous of all the ways a country had of asserting itself. And the German historian, Ranke, inherited this tradition – he was not responsible for its creation, and it sprang not from a sort of Prussianism, but from the University that was closest to England. Ranke was of higher intellect than his predecessors, however, and he was better able than they to see the intellectual implications of the whole system. He put forward a doctrine that became famous in the 19th century and was widely discussed again after the Second World War: the doctrine of the *Primacy of Foreign Policy*.[4]

The doctrine itself was rather uncongenial to Englishmen for a period of about a century; but Ranke would have understood this and would have predicted it. One aspect of the theory would be that, prior to the domination of airpower, an island state need not give the predominance to foreign policy – need not even give its perpetual attention to foreign relations. An island state was so free from the most urgent kind of pressure on that side, that it could give its primary attention to internal politics and its own constitutional development without ever realising that a factor in foreign policy – a factor in geo-politics – was responsible for its happy position. But German historians in the Ranke school have often pointed out that if your state is bounded by land frontiers of a very emphatic kind – if you have a neighbour like Russia on top of you all the time – you are never quite released from fear; foreign policy does become a constant preoccupation and you can never have a very relaxed attitude to the problem of war. Also, your foreign policy predicament affects even the kind of constitution that your country is likely to develop. Britain, because she is an island, could afford to develop liberty, have party divisions, and put checks on the executive. A frontier state needed a stronger arm of government, a more united front; for the monarchs who most provoked the kind of rebellions that led to the enlargement of liberty were those who failed in foreign policy – King John, Charles I, perhaps James II and George III. Ranke himself wrote a *History of England in the 17th Century* and used diplomatic documents – showing the importance of foreign policy in the story. A pupil of Ranke's, Von Sybel, wrote a history of the French Revolution in many volumes,[5] and one of the results of his work was to show the importance of foreign policy and war in the development of the Revolution itself[6] – a point taken up later by the famous French historian Sorel in a work of many volumes.[7] The doctrine of the Primacy of Foreign Policy may have been exaggerated somewhat or even perverted by some of the disciples of Ranke; and in fact it did not prevent Ranke himself from being greatly interested in cultural history, for example. But it is not a doctrine to be summarily dismissed and now that the navy is no longer a secure shelter for England, the fear of an over powerful Russia has become as real to us as it was to Germany before 1914.[8] And it would be difficult to calculate the degree to which the fear of Russia during something like twenty years has affected men's permanent attitudes, the character of culture, the course even of scientific development and the modern acquiescence in what I might call the high-powered state.

England, then, was slower than some European countries in realizing the importance of foreign policy. She was slower also in developing her diplomatic studies. When in the 1890s Oscar Browning began studying English diplomatic history in the time of the French revolution and started publishing some of the foreign office papers, he commented on our slowness and pointed out that foreign historians had worked more on our diplomatic papers than we had. When Holland-Rose published his big life of Napoleon in the early years of the twentieth century,[9] he too pointed out that foreign historians had both studied and published the English documents that we ourselves had been neglecting. Next there came Harold Temperley and Charles Webster who made big studies of Canning and Castlereagh, carrying the development into the post Napoleonic era, the age of Metternich. They were still working in these fields when I began my career as an historian and they established schools of historical research in this field, their pupils carrying their enquiries gradually further and further into the 19th century. When I was young, it was the fashionable thing for research students to receive their initial training in some field of diplomatic history; and this was understandable – as I have said, the documents were all tidy in the archives, beautifully arranged and easy to handle. The war of 1914 and the peace treaty of 1919 had convinced many people of the importance of getting training in the technique of foreign policy. You really felt that you were getting at the secrets of state – for there was more concealment in foreign policy than in most things; you could be pretty sure of bringing something new to light. Furthermore, though you might be studying documents in the British archives you would be studying our relations with Austria or with Prussia or with Russia or with France – you would be examining another country's foreign policy too – and you would be checking the story in the British Foreign Office papers against the story as it appeared in the archives of Vienna. The English point of view would differ from the Austrian, and the English account of the whole story would differ, so that you had to collate the two and advance critically towards the real truth – which required detective work and imagination, and a close collation of English records with foreign sources. And in fact diplomatic history was a good training for research students – it was very precise, you could follow the story day by day, and it was a story in which something very big might have depended on quite a microscopic event. I have sometimes said that even if one intended to move later into a different kind of history – the history of thought for example – as I myself did – there was a strong case for having one's initial training in a field so precise, so scientific, as that of diplomat history.

And so things went on – the research students gradually moving their studies forward in the 19th century – until by the middle 1930s, things were getting congested and we began to wonder what was going to happen next. If you look either at published books or at articles in learned journals, you will be amazed at the number of students in England and America who were working on the Balkan question in the period 1875 – 77. The situation was becoming serious – for the British archives were only open to the year 1878. It looked as though it was going to be difficult for researchers to find a new subject, a piece of un-trodden ground. And this was the first of the reasons why, in the middle of the 1930s, there began to be a reaction against diplomatic history at the research student's level. Even at the present day, our archives are only open for the period down to 50 years ago; and it is not possible for the ordinary student to see the archives for the study of the origins of the war of 1914,[10] though single historians may have special permission to see certain things. One of the points that historians are continually taking up with government is the unsatisfactoriness of this fifty-year rule. And, of course, the very fact that governments feel the necessity of keeping the documents secret for so long, demonstrates that there is something to hide, something that the government or its agents don't want the historian to find out. Sometimes papers are released to what we call official historians, who have partly been chosen because they are regarded in governing circles as safe, and who are put under oath not to reveal anything without due authorisation or without a chance for the censor to stop them. It would generally be held that a subject is not properly released to the historian, unless the relevant documents are made available to historical scholarship without any limitations.

(*Editors' note*: Here the discussion of diplomatic history ends suddenly and continues with desultory observations on the creative role of diplomacy, its ability to expand political vision and perspective, and its civilizing function in the face of naked power – recurrent themes in Butterfield's other writings. In an earlier draft, however, similarly titled, he elaborates further on the inter-war marginalization of diplomatic history, and it is with this transposition that the text continues below.)

But there were more important reasons why diplomatic history went out of fashion and was openly attacked in the 1930s; and this was a thing that operated not in the higher levels at universities but rather in public opinion and the world at large. The First World War had one curious result for which I believe there was no parallel after the end of the Second World War. It generated a species of utopianism, a general consensus that, the Germans being now disposed of, a new world had come into being – a world of peace and of ever expanding democracy – the future was to be free from war. In fact, it was thought that wars had been the result of monarchical regimes, so many of which were now being liquidated, and the real instrument by which they had been brought about was a thing called the 'old diplomacy.' Of course the 'old diplomacy' hadn't succeeded in preventing all wars, but the current argument did less than justice to what had been achieved in oiling the wheels of international relations, staving off conflicts and minimizing wars. If Russia, the United States, and Great Britain met in conference today but failed to solve the problem of Berlin, we should not argue that the conference system was responsible for the difficulty; we should merely say that it had failed to solve it. But after 1919, diplomacy became the scapegoat, as though, if you could get rid of it, there would be nothing to prevent states from being friends. The total result was tantamount to turning your back on the previous experience of the world in the field of international relations – a field in which it takes a lot of experience to learn all the snags. And in the inter-war period we committed some howlers which our predecessors had learned to avoid. All this involved a deeper error – almost a philosophical one – of a kind, which for obvious reasons, there was no danger of our adopting after the Second World War. The past was deplored as presenting a vast panorama of power politics; and instead of studying this demon, this enemy, for the purpose of mastering it; you turned your back on it – you wouldn't take part in the game of 'power politics,' as though by closing your eyes to the fact, you could somehow conjure the bogey away. And I have known people who said, 'But the power-politics are patently there,' and they were pilloried. It was assumed that they must like that sort of thing – when what they wanted to do was to face squarely the magnitude of the problem that they were confronted with. I think it is probably true to say that both English and German intellectuals – in vastly different ways – have gone wrong in their appreciation of the role of force in human history; but we went most wildly wrong in the period between the two wars, when many people both wanted us to put a check to Hitler and wanted the disarmament of our armies and fleets. I suppose that by now a younger generation cannot even imagine the appalling fears provoked in Western Europe by the march of Stalin in the latter half of the 1940s, before the United States had pulled itself together – the awful shadow that was produced in the minds of men by the proximity of overwhelming power. The problem of how to deal with a world of nations in which that kind of thing can occur, is one of the greatest ever presented to the human intellect and will be with us long after we have surmounted the difficulties of getting to the moon. We can't afford to turn our backs on the centuries of the world's experience of these problems, or to lose sight of any trick or device that great men discovered in the past. This problem of power is a fundamental one, and both the structure and texture of our history are going to be seriously wrong if we have not come to grips with it. Englishmen used to taunt Hitler because he chose to go in for guns rather than butter. I personally prefer the butter and as I am a near pacifist, I don't care about the guns – I don't care about being in a position to exert power or outdo anybody else in power. But the thing we forgot in the 1930's was a sense in which predominant power is unanswerable at least in mundane[11] terms. The man who has the guns has a good chance of getting the butter too – and that is the problem diplomacy tries to cope with.

In the latter half of the 1930's, there was a further movement, definitely directed this time, against diplomatic history as a subject of study. These were the days when the opposition to Hitler created a certain sympathy with Hitler's *bête noire*, the Communist ideal, and people opened their minds to the Marxian interpretation of history to a degree never paralleled in England either before or since. And it happened that Marxist history had greater effect through the indirect influence that it exercised on non-Marxists than through the people who were converted to the entire system. It came to be widely held that the economic factor, and even the economic motive were supreme in the realm of foreign affairs and war; and it was inferred that they had always been supreme. If you studied the economic history, therefore, you were getting the essence of the matter. There was no need to burden oneself with the diplomatic narrative.

And the odd thing was: that in the latter half of the 1930s, public opinion began to take an interest in international affairs, and to assert its will in foreign policy, more than ever before, or ever since. Democracy claimed a voice in the detailed diplomatic moves that the government made. And, at the same time, opinion set itself against diplomatic history.

The problem at the stage to which I have now brought it, brings us up against a hurdle which I can only meet by facing an issue that would be very controversial in the historical world today. It is the issue of history as a study of general movements in society or the world at large versus history as the study of government action. And, here, I should want to qualify the views I have sometimes expressed in print. I once tried to describe how (and why) in the 19th century historians concentrated so much on governmental action; so often studying events or changes in the world from the point of view of government and how nowadays we had become aware of the important movements that rise spontaneously in a country, acquiring a momentum of their own.[12] We had our eyes on a kind of history making that just proceeds out of men living in a society – the government acting only on the margin of this. I don't want to deny the truth of all this; but when you look at long stretches of the past and try to see the story as a whole, I think there is an important footnote to be made to it – a general comment on the way history works. Even supposing you take your start from the view that I have just described, and decided that you will study society as a whole – supposing you begin by assuming that the history of England is really produced by the activity of the whole of the English people scattered over the length and breadth of the landscape, you are brought back at the finish to a heightened view of the importance of government action. Whether the conservative party stays in power or goes out next year, the curve of development in England as a whole is going to be affected very much by influence of the party that captures the government. In other words, the person who captures the reins of government does acquire extraordinary leverage in the historical process. The 19th century imagined that when Europe became divided into something like our nation states – their shape on the map coming to be fairly recognisable in the 10th century – the underlying cause was a spirit of nationality in the peoples themselves, their differences in language, and so forth. We now know that the dynasties were the real creators of this pattern on the map – the various countries that crystallised are represented.

Some forms of history are more technical than others; and perhaps the highest kind is that which is more than mere technique – the final sort of history, which, after everything has been divided up for scientific analysis, recombines the parts and looks at the story as a whole, without the separation of economic, constitutional, diplomatic etc. However, before this can be reached, there have to be the professional analyses, applying their techniques to all the parts, and if history is to be a university subject, a serious branch of scholarship – not merely a thing to read books about but a study on a level with the natural sciences – shall we say a study leading to a degree – we have to expect it to involve this tougher discipline, which implies a certain training in technique. Of the various kinds of this more technical history, the constitutional has for a long time been by far the strongest in this and other universities (and I do not know to what degree this has been the cause or the result of this entrenched position that it secured as a vested interest) I suppose it established itself at a time when the development of our political institutions and the mechanics of self-government were the overwhelming preoccupation of Englishmen. I wonder whether, nowadays, our preoccupation is not really rather international so that foreign affairs and even the practice of diplomacy have become more relevant for us.

We today need no reminder about the unmistakable existence and formidable character of power-politics, and diplomacy functions where there are real pressures and tensions, because the element of power is so grimly involved. It was one of the crucial mistakes of the interwar period to imagine that diplomacy was a mere matter of reasonable men discussing theoretical rights – something like a study circle or roundtable discussion – like a collection of dons in Cambridge debating from first principles who should own the Antarctica. Diplomacy has its *raison d'être* because it is the wills of men that are in conflict (it is not merely their minds that are engaged in argumentation); and because wills are involved, the element of power is involved – indeed if there were no resort to diplomatic negotiation it is understood that there would be nothing left except the appeal to force. Even where negotiation exists, it does not get rid of the power, but rather makes it change shape and operate through a different channel – it operates in a more refined state and a different temperature. Just as the prime minister can more easily borrow a pound than a beggar, so the reasonableness of the United States can normally operate to a greater effect than the same degree of reasonableness in Belgium if the two are engaged in trying to settle a conflict over a territorial claim. It has often been forgotten in the 20th century that nations sometimes arm themselves not merely for direct defence and not merely because they want a war, or intend one, but purely in order to have some 'pull' in negotiations, so that their voice will at least command some *respect.*[13] Of course if you have a powerful ally in the background, that itself will increase the pull and will make the trial of strength one stage less crude, one stage more indirect. If you put your case well you can give the other party reason to pause, or show him that it will profit him to be reasonable – you might get moral support from outside powers. Indeed, if you are a clever diplomat, your effectiveness in negotiation may be out of proportion to any basic military strength you have. And when you are involved in this kind of transaction, it will be necessary at crucial moments to be doing a piece of mental arithmetic – calculating the forces and the chances, in order to see how far you can go – in the way that you might calculate that if you tried to borrow £500 from a cousin of yours, he would be likely to think you distressed, unlikely ever to return the loan, but if you touched for a mere fiver, he might lend it with greater confidence.

All this, of course, requires experience, and the study of the history of diplomacy is relevant not only because it gives you information about the position, the interests, the tendencies, the policies of various states, but also because it can add the benefits of what I should call indirect experience. But it is one of the cases where, if you have no experience to start from, history is liable to be so deceptive – because once the story is over and you look on it afterwards, everything seems so easy – just as hitting a ball for six might look easy to someone who had never tried it. In fact, that is one of the pitfalls of history – the discoveries of Sir Isaac Newton look easy enough even to a schoolboy nowadays – and the actions of statesmen look simple because, unless you study them in great detail, you don't see the man quivering on the brink of a decision; rather like a batsman deciding on some brilliant stroke as the ball comes towards him. I sometimes feel like saying that I have known Napoleon and Bismarck shown up as boobs in foreign policy by students who wouldn't have had the diplomacy to wheedle a pound out of their college porter.

I need not say that it requires a great deal of diplomacy to climb to the front place in the government of one's own country; but it needs rarer gifts than to become the predominant statesman on a European continent where there exists quite a handful of considerable powers. Metternich is often described as a merely silly man, but he is one of the very few who managed to achieve such a position. I believe that in modern times you could count on the fingers of one hand the men who attained anything comparable with it. Hardly anybody held the position as long as he did – yet he did all this at a time when Austria did not posses a military power[14] that was commensurate with it. Therefore, diplomacy is a creative thing – it is capable of producing masterpieces – and sometimes it has prevented wars when they looked more inevitable than that of 1914 seemed to be. It is not always realised to what a degree, Louis XIV used diplomacy in order to minimise the actual amount of war that he should have to fight in his aggressive period – how often he was able to use diplomacy instead of actual fighting. There have been states which, by good diplomacy, have turned their actual weakness into a source of bargaining power; have used it to give them leverage in negotiations. I wonder if I shall be understood if I say that diplomacy enables men not to eliminate the power factor from international politics but to steal a march on this beast; steal a march on naked power. It represents the encroachment of human reason upon what would otherwise be a jungle. And if you condemn it because in it, sweet reasonableness is still mixed with deadly power, you are forgetting that the alternative is the release of the world to brute force.[15]

What is amazing, in point of fact, is the degree to which diplomacy comes to civilise itself, the velvet getting thicker and thicker over the steel, so that a world of tremendous politeness and urbanity seems to evolve if nations are on fairly reasonable terms for a considerable period. They used to talk about ambassadors 'lying' abroad for the good of their country,[16] but I am astonished at the degree to which, in the 19th century, an ambassador who was a liar seemed to be a marked man.[17] Of course, if diplomacy were perfect and all the wheels were properly oiled and there were less chanciness in life, there would be little need for war – you would do the mental arithmetic, calculate the forces, measure the chances and know whether it was the prudent thing to give way. But the texture of the time process is such that people like to make gambles.

In all this, I am talking about the diplomacy of historical Europe. If I were to make adjustments for the present day, I would guess that there is (or there was until recently) more of the 'old diplomacy' at UNO than many people realise. Yet, if one looks at the ordinary diplomacy as conducted by the various Foreign Offices, I believe that what might be called 'the moral factor,' has been much more important there, particularly in the last decade or so – partly, no doubt, because both East and West, have been bidding for the moral support of the uncommitted parts of the globe.

I always think that there is a deposit of experience which the world accumulates as the centuries pass, and those who read diplomatic history in a way that brings the issues really home to themselves, can get a better feeling for the way things work in diplomacy, or, from the writings of old practitioners of the art, they can learn the snags, acquire the maxims, realise the paradoxes. It would be only a personal view, but still my view that modern democracies, particularly down to 1945, lost something by cutting themselves off from the tradition. One of the things we forgot was that in a world of armed powers you must not expect your word or your will to carry the same weight if you are disarmed as if you are armed. Another of the things we forgot was the very considerable repertoire of maxims which had been evolved to the effect that if war did break out, you should make it a war for limited objects – prevent it from becoming anything like a war for religion. If you are fighting for oil in Persia, say that you are fighting for oil in Persia, because then you can negotiate; you can come to a compromise – don't say you are fighting a war for righteousness,[18] where you merely mean that the other power is unrighteous in claiming the oil – for then it becomes a war for the destruction of the righteousness and compromise itself becomes wicked. It was an old maxim that you must use foresight – that it is wrong to wait until an aggressor has actually emerged in full power before you try to check him; the way to stop an aggressor is to check him before he has had a chance to show his hand, before he has reached a position in which he can act with impunity. On the other hand, it is unwise to let a power become so desperate that it will pay any price to secure escape or revenge,[19] saying 'very well, we must go under but we can pull everything else down with us.' It is wrong when fighting an enemy to forget entirely that at the next stage in the story you may need that enemy as an ally. And it is wrong even to wipe out a state, to destroy a great power – since the power vacuum that you thereby create will conjure into existence another bogey worse than the first. In the 20th century I believe, we have broken down once or twice at the point where a treaty needed revisions, but all the signatories had to agree to this, and the power to whose detriment the revision was to be made possessed therefore a *liberum veto*. I think I am right in saying that England in the 19th century found a way out of this locked circle. At any rate, I suppose that many of us have reason to be aware of one important maxim: that you cannot have real peace until you have devised peaceful means of getting out of that circle – a peaceful way of revising the *status quo*.

Here, as in politics generally, it is a mistake to let wisdom solidify into hard and inflexible maxims, and it would be best to regard these only as pointers to the essential thing; the acquisition of a sort of experience, a proper feeling for the way in which things work.[20] But there is a further sense in which diplomatic history carries us outside itself and that is by its expansion into what almost might be called the world of Theory. In the opened documents of foreign offices, but still more perhaps in the publicistic writings issued on behalf of government, or by independent observers, the diplomatic story, which is always one of day-to-day detail, gets translated into types of doctrine. One example would be the doctrine of the balance of power which occupied so important a place in the literature of this subject, particularly in the epoch of Louis XIV's aggressions and in the following 150 years.

Here was evolved the view that in an international world which in the last resort was based on force, only the balance of power – a peculiarly European, peculiarly modern concept – permits the existence of small states which are more than satellites, and secures for small states a significant and independent role. Behind all this, however, is the larger theory of an international order, in which all states are regarded as, so to speak, members of the same club, jostling and squabbling to acquire the best chairs or the best positions in the club premises yet all (if they are farsighted about their own interests) concerned for the preservation of the club itself. Nobody has been able to prevent Foreign offices from being egotistical in the sense that they can hardly help looking upon it as their function to take care of the interests of their own country. Yet there emerges the view that, even in their long term self-interest they must work for something higher at the same time; and this something has not to be Catholicism against Protestantism, or democracy as against monarchy, or Liberalism against Socialism. You do not work for a Catholic international order but for just an international order as such – one which can comprise Protestants and Catholics, Republicans and Monarchies, Socialism and Liberalism. And as the international order grows, developing into this club, diplomacy itself can become more civilised and urbane.

Notes

1. This essay was originally delivered as an address to the Cambridge University History Club, November 1963 (eds).
2. Cf. H. Butterfield, *The Whig Interpretation of History* (London 1931 1st edn) (eds)
3. For the text of these see: E. Alberi (ed.) *Relazione degli ambasciatori veneti al senato* (15 vols. Florence 1839 – 63) (eds).
4. Italics by Butterfield (eds).
5. *History of the French Revolution*, trans. Walter C. Perry (London, 1867 – 69) 4 vols (eds).
6. Cf George Iggers. *The German Conception of History* (Middletown, CT 1983) chs 4 and 5 (eds).
7. *L'Europe et la Revolution française* (Paris, 1885 – 1904) 8 vols (eds).
8. It must be remembered that Butterfield was writing before 1989 (eds).
9. J. Holland-Rose, *The Life of Napoleon I* (London, 1902) numerous subsequent editions (eds).
10. This now, of course, is no longer the case (eds).
11. Italics by Butterfield (eds).
12. Cf. H. Butterfield, 'Tendencies in Historical Study in England,' *Irish Historical Studies*, vol. IV, no. 15. (March 1945) pp. 209 – 23. idem, *The Study of Modern History: An Inaugural Lecture* (Cambridge, 1944) (eds).
13. Italics by Butterfield (eds).

14. Underlined in original manuscript (eds).
15. Underlined in original manuscript (eds).
16. The reference is to Sir Henry Wotton, who while on his way to serve as English ambassador to Venice in 1604, allegedly coined the wry epigram 'Legatus est vir bonus peregue missus ad mentiendum respublica causa (i.e. 'The envoy is an honest man sent abroad to lie on behalf of his country') cf. G. Mattingly, *Renaissance Diplomacy* (New York, 1955) p. 301 note 7 (eds).
17. Underlined in original manuscript (eds).
18. Underlined in original manuscript (eds).
19. Underlined in original manuscript (eds).
20. Underlined in original manuscript (eds).

3
Prussia's Attempt To Make Separate Peace with Napoleon After Jena

The fine show of armies arrayed and kings in proud posture had collapsed into medley and muddle in the twin battles of Jena and Auerstädt. It littered the countryside and strove painfully to sort itself out afresh, as the din subsided and the smoke cleared away. Napoleon, sprawling upon the ground to study his maps, fell into an undignified sleep, and his guard took silent stand around him. Murat, writing to his master, apologised, 'Your Majesty will pardon my scribble, but I am alone and dropping from fatigue,'[1] and he who was no weakling blundered the strokes of his pen, as if he had been drunk. The release from the tension of battle liberated the worst elements in an army. The forces of the very victors seemed endangered by indiscipline.

> 'The footsteps of the corps d'armée', wrote Soult a few days later, 'are marked by fire, devastation, and crimes atrocious beyond belief. The orders of the leaders are despised, the lives of officers are endangered, and as a crowning evil the resources and food afforded by the country are destroyed as French troops come upon the scene.'

A stray English traveller who came by soon after the battle found a devastating sight, the ground littered with papers, letters, account books, pamphlets and miscellaneous unwanted goods, while the French rummaged baggage waggons, tossed part of the booty to the country people clamouring around, and with the recklessness of light-hearted soldiery invited all passers-by to stop and share their luck.[2] All was in disarray and men went off to shoot game or to steal poultry, to ransack houses or to pilfer from farms. 'Never was pillage carried further than in this campaign and disorder extended even to insubordination.' The Prussians, in still worse condition, had all their evils complicated by panic. In the morning they had lightened their burdens by throwing away their food, and the end of the battle found them scattering through the villages to satisfy hunger and thirst, digging up potatoes and turnips in the fields, and rushing blindly, officers and men alike, ignorant of the true direction for retreat. An English minister, Lord Morpeth, took to his heels, abandoned his mission before it had really begun, and found a speedy vessel for England. Haugwitz, who had the direction of Prussian policy, made hurried escape to Magdeburg. The king himself had great difficulty in evading capture, and only after many miles of riding hither and thither, many hours in the saddle without food of any sort, did he find his way in the face of the French watch fires and through the midst of his own straggling forces, to a brief day of rest at Sömmerda. On his way he saw the two streams of fugitives from Jena and Auerstädt meet and cross in confusion; and everywhere were tumults and alarms, fugitives and panic-stricken herds, disorders and wild rumours, and an infectious atmosphere of despair. The queen had left her husband on the eve of the disaster and was making her way to Berlin. Stories of battle followed her, reports began to be spread concerning a great defeat of Napoleon, it was even whispered that Murat was dead; but when she was but a few hours' journey from Berlin the truth overtook her and a little later she received a sorry letter from the king. She found Berlin in a fever; her children had already been sent to Stettin and it soon became apparent that she must continue her journey, merging it now into a flight for existence. She wrote to the king: 'Everywhere there is the hope that all is not yet lost and that God will help us. You still have troops and the people adore you and are ready to do anything.' But the king, Frederick William, had never had any heart in this conflict, and it was noticed that he now showed a sort of bitter satisfaction at having found his own premonitions come true. Even before the battle there had been a heart-breaking series of war councils at Erfurt, there had been distraction and delay; and now that the battle was over it was easy for the military men who were at hand – men like General Zastrow and General Köckritz – to whisper the right word in his ear and induce him to ask for an armistice.

The war had lasted only a little over a week. The Prussians had gone into it with flags flying. All the wild enthusiasms, the bragging pride, the rattling swords, had sounded the din of a confident untried nation newly entering upon war; and the memory of Frederick the Great had lured these people on. It was on the 5 October 1806 that the Prussian minister at the court of the Emperor Napoleon had handed an ultimatum to the foreign minister Talleyrand, and the ultimatum had expired on the 8 October. On the 12 October, after hostilities had commenced, Napoleon had made a final appeal for peace and had asked the Prussians to prevent the great battle that seemed impending; but the Prussian king claimed that the letter did not reach him in time, and Jena was fought on the 14th. Frederick William's demand for a suspension of arms came as his answer to this final overture from Napoleon. With it was coupled a regret that the pacific proposal had arrived too late to prevent the carnage. In this way, at one touch from Napoleon, Prussia crumpled, and cringed, and came whining back to the side of France.

Napoleon refused to grant an armistice. His reply to the request betrays his preoccupations in this October of 1806.

> Any suspension of arms which would give time for the Russian armies, whom you seem to have appealed to in the winter, to arrive, would be too inimical to my interests for me to be able to subscribe to it...I do not fear the Russian forces.... I met them in the last campaign. But Your Majesty will have more to complain of than I. Half of your States will be the theatre of war.... The other part will be ravaged by your allies and will suffer still more. It will be an eternal subject of regret to me that two nations which for so many reasons were meant to be friends, have been drawn into a conflict so purposeless.... Still I must repeat to Your Majesty that I shall be glad to see a way, if it is possible, of re-establishing the ancient confidence which reigned between us, and of reconciling the sentiments that I bear towards you with my duty and with the security of my people, compromised once again by this, the fourth coalition in years.[3]

Napoleon had been fighting this Prussian war, in fact, with his eyes fixed upon a corollary, and that corollary was the entry of the Russians into Poland. It was a contingency he had merely divined, but if it should occur at all, he would owe Prussia a grudge for it. So in his reply to their demand for an armistice he turned even the approach of Russian friends into an added terror for these Prussians that they might fear it more than he himself did, and so end by trying to prevent it. Frederick William misread the conclusion of this letter, thinking that he saw in it a promise of kind treatment and quick forgiveness. In reality it was a pointed and deliberate and ominous expression of a doubt. Napoleon was not sure whether policy and events would leave room for the good sentiments that he still affected to have for the Prussian king. He was not sure that he could afford to be friendly to the court of Berlin at all if the Russians brought war into Poland.

At Magdeburg the King of Prussia met Haugwitz who had charge of foreign policy and had led the country into war; also Lucchesini who had been ambassador at Paris, and the secret-cabinet councillor Beyme. They were depressed to hear of the refusal of the armistice and they discussed the course that Prussia ought now to pursue. They all agreed that it was necessary to appeal for peace. Beyme wrote an alarming letter to his wife in Berlin, in which he declared that 'the situation was such that one must not bid but beg for peace; it would be presumption to make a proposition; one must surrender completely'.[4] The words were spread abroad; it was no secret that negotiations were being entered upon; and everywhere arose an expectation of early peace, a certainty that the treaty would be an ignoble one for Prussia, a feeling that it was useless to go on fighting while the king and his ministers were surrendering everything in abject negotiation. One can understand why, in such an atmosphere, Prussia offered little resistance to the French.

Lucchesini was chosen to conduct the discussions with Napoleon, and he received full powers to conclude a treaty. He was instructed to consent to the cession of Prussia's Westphalian provinces up to the Weser, to agree to the sacrifice of Bayreuth, but to use every effort for the securing of an armistice. He was not anxious for the odious mission, and wrote:

> On the morning of the 18th [of October] the king and Count Haugwitz forced upon me the wretched commission of going to ask the French Emperor for peace. I could not get out of this thankless and mournful task. I was charged to draw up my own instructions and to write the letters to the Emperor myself. The king approved and signed them and then left for Custrin. I left Magdeburg at noon on the 18th.[5]

On the 20th he reached the French armies and dispatched the king's letter to Napoleon, but was himself detained at the outposts. 'I have sent Duroc to see what he wants,' Napoleon wrote soon afterwards, 'and I am waiting for his return. The Prussian king seems quite decided to come to terms. I will do this, but I shall not let it prevent me from entering Berlin'.[6] He appointed Duroc to treat with Lucchesini and at midnight on the 21 October the negotiation for peace began.

The Prussians were not wise in their choice of a negotiator. Lucchesini, when ambassador in Paris, had more than once proved an obstacle to the smooth working of the Franco-Prussian alliance, and long before the emergence of the controversies which had led to war, Napoleon, who disliked and distrusted him, had talked of having him recalled. He had been responsible not for creating but for exaggerating the fears and distresses that had precipitated the recent struggle, for he had not only transmitted some wild rumours to Berlin, he had advised his king to take carriage and fly direct to St Petersburg; and Napoleon had been to this extent correct in charging him with keeping bad company, listening to wild talk, taking his information from speculators, and writing dispatches that made trouble. From that moment Napoleon had abused his character, and had complained to Berlin of his unfitness, until the man had been glad to be recalled. It was not tactful of the Prussians, when they were seeking the best escape possible from the predicament into which the war had brought them, to send for the purpose the very man whom Napoleon had seised upon as the prime mover of the quarrel, the cause of all the trouble.

Napoleon was more careful in his choice of a negotiator, and the appointment of Duroc was significant. It was necessary to prevent Talleyrand from having a hand in the discussions; he might juggle with the negotiation and insinuate his own ideas into the affair, or try by devious processes to smooth down the severities of his master. No diplomat, but a tool, a cat's-paw, the secretary of Napoleon, was chosen to treat with Lucchesini. He was still to be mere secretary, nothing more than a scribe, the unprotesting servant of his master's whims. Napoleon loved Duroc for his faithfulness in this kind of service. 'His cold and unexcursive mind,' wrote Bourrienne, 'suited

Napoleon, whose confidence he enjoyed until his death.' General Rapp has summed up those qualities in him which made him particularly useful to his master in the negotiation of the peace with Prussia:

> Duroc was a man whose services were almost indispensable to the Emperor. He always enjoyed the highest favour and the greatest confidence, which he in every respect deserved. Few men were so distinguished for tact, spirit of business, and skill, as Duroc; and at the same time few were so remarkable for modesty. His devotion to the Emperor was without bounds. He had a good heart and was an honest man; his only fault was his fear of displeasing and his excessive timidity.[7]

Napoleon in appointing this man to treat with the Prussians really kept the whole negotiation entirely in his own hands.

At midnight on the 21 October Duroc and Lucchesini began a conference that lasted for a few hours and discussed the terms of the proposed peace treaty. The Prussian plenipotentiary spent the rest of the night in thinking out a reply to the proposals that had been put before him; ready for the second interview that took place on the 22 October he was kept moving about from place to place as the French armies advanced, but Napoleon refused to see him personally. On the 23 October, at Wittenberg, a more official conference was held, and on the following day Lucchesini was informed of what Napoleon's intentions were. He was given the basis of the proposed treaty of peace. On the morrow again he was in Potsdam and wrote to his master. The terms that had been offered him were so exacting and costly that he would not take it upon himself to sign them.

Napoleon's conduct in the meantime had given the Prussians little reason to expect an easy conclusion of peace. The French armies were sweeping across the country and into the four corners of Germany to hunt out the remnants of Prussian military strength. The battle of Jena had been made the starting-point for the complete subjugation of the defeated state under a system of military occupation and financial exaction. Napoleon entered Berlin in triumph. He seems to have been intoxicated by the very act of fighting, and the exhilaration of conquering so easily. One would have inferred that he had no motive left for ever desiring the Prussians as his friends, no political object in view to make him wish to soften their humiliation. He seemed like a man with a great resentment, as though this war, for all its laurels and victories, had cut across his path and had intervened to spoil his plans.

So while Lucchesini had been engaged in flying conferences, negotiating peace with Duroc but at the same time moving from town to town to keep pace with the advance of the French armies, crowds of fugitive Prussians, a little ahead of the invaders, were being driven into exile and were flying to escape capture. From Berlin the wealthy, the noble, the official classes had taken to flight on hearing of the battle of Jena. All who had the means had hastened away; all the booksellers, it was noticed, seemed to remember Palm; and it was said that on the afternoon of the 17 October not a vehicle was left in the city, not a horse, not even a donkey. The military governor of Berlin was among the refugees; the government was dispersed, the king and queen were known to be in flight somewhere, and were trying to meet one another. The journeys of these Prussians were suffered under the most nerve-racking conditions. A description exists of how from Berlin the travelling carriages were pursued by the hootings of scoundrels and the menaces of the dregs of the populace who did not like to see themselves so deserted, and hastened back to scrape at the heels of the French. Sometimes the road would be blocked with fugitives; fear drove them on, bad news followed them round; mishaps did not fail to occur on journeys so taken under pressure. Horses would be driven to the last stages of exhaustion and fatigue; the breaking of a carriage wheel would come to overwrought minds as a nerve-racking catastrophe and would hold up a procession of miserable refugees; and all the attendant circumstances were calculated to induce dejection. If a pause were made, and if at last there were some feeling that the flight might be at an end, still the sound of cannon, or the rumour that troops had been seen across the river, would start the race again. It was in such an atmosphere as this, and not in cool council chambers, that the king had to decide important questions of policy. The court was in full flight. The ministers were dispersed and overtaken with panic. Conferences would be driven from town to town, held so to speak under the cannon's mouth. It was natural that these people should judge the situation of Prussia from their own immediate circumstances. It was difficult for them to meet their political problem with a mighty exercise of will, when every petty detail of their daily life was conspiring to keep them in dejection. It is easy to understand why they looked to Lucchesini, and put their hope in the clemency of Napoleon, and set their hearts on the conclusion of peace.

Frederick William himself, waiting for news from his minister Lucchesini, was in one of those moods when a queer psychoanalytical apathy fell upon his repressed mind, and he would busy himself with odds and ends of futile occupations and interests, like a harassed man who picks at a piece of string. He, who was 'rather the conscientious drudge of state than its first servant', could not bear the awfulness of responsibility and shrank from taking a decision in matters he knew to be so momentous. From his way of walking around, from his habit of toying with trifles, people said he was indifferent – that he might have had ten kingdoms to lose. Blindly he clung to his old ministers, Haugwitz, Zastrow and Beyme, keeping the more vigorous Hardenberg out of his counsels, lest he should hinder the conclusion of peace. The queen herself was aware that it was her own policy of war with France that had come into discredit: she might be anxious to continue the struggle but she would not interfere in politics any more. On the 20 October she joined the king at Cüstria where the inhabitants of the surrounding country were already rushing within the fortified walls, dragging everything they could carry. All the hope of the ministers was centred in the negotiation that Lucchesini was conducting; but there was certain nervousness, since Lucchesini had been keeping silent for a time. Frederick William wrote a pitiable letter to Napoleon, expressing his anxiety and complaining of the delay; and this letter seems to have crossed with Lucchesini's dispatch announcing the terms which Napoleon had dictated.[8]

Lucchesini's report now reached the Prussian king, and the Prussians were at last able to know the kind of treatment Napoleon intended giving them. The peace terms that were now presented were the ones that Lucchesini had not dared to undertake the responsibility of accepting, and they showed Napoleon more severe than the Prussians had contemplated. Prussia, according to these terms, was to renounce the provinces she had hitherto possessed between the Rhine and the Elbe and to abandon her former relations with Germany. She was to engage not to interfere in the affairs of Germany but to promise that she would recognise the drastic changes that Napoleon contemplated there, particularly the great extension of his Confederation of the Rhine. She was allowed to retain the Duchy of Magdeburg and the old Mark of Brandenburg, so that as a result of the territorial rearrangements her centre of gravity was to be shifted eastwards; in future she was not to allow preoccupation with German affairs to obscure what Napoleon regarded as her true function in Europe – she was to concentrate upon her role of buttress against the powers of eastern Europe, and cease to be the competitor of Napoleon in Germany. She was given a geography that would itself set the conditions of her future policy and turn her attention eastwards. This treaty was Napoleon's answer to the challenge that the court of Berlin had given to him in the year 1806. He had wanted to keep Prussia as an ally, as a client state; she had been useful to him because she could check or frighten Russia; but she had chosen to adopt rather the role of a German power, she had entered into competition with him for that hegemony of Germany which Austria had allowed to slip from her hands, she had been willing to leave him his influence in south Germany, his Confederation of the Rhine, but she had come to issue with him over the question of the leadership of the north. This being the nature of the struggle, and Napoleon having won in any case at Jena what he might consider as a special title to the overlordship of Germany, it was natural that the guiding principle of the peace treaty should be the final settlement of the problem of Prussia's place on the map – that is to say, the exclusion of Prussia from the Germanic system.[9]

On these terms a Prussia greatly weakened, and so unable to imagine herself a possible rival to France, was to pay a huge indemnity for the war she had brought about, and was to return to her place in the French system of politics; she was to cooperate with Napoleon in the war against England by closing her ports to British vessels, and she was to guarantee the integrity of the Ottoman Porte, agreeing to join France in its defence if the Russians should attack it. This last clause reveals the context of the whole treaty and gives at least a clue to the situation from which it proceeds. There was no mention of what Prussia must do if the Czar turned his forces into Poland. That contingency was not contemplated for the moment; in truth it had never been more than a wild surmise of Napoleon's, and he was now beginning to doubt it and to change his ground a little. One of the bulletins of the French army at the time when these terms of peace were communicated to Lucchesini, stated: 'We have not yet received news of the conclusion of a treaty between Prussia and Russia, and it is certain that no Russian troops have appeared to this day on Prussian soil.' Later, on the 18 November, when it had become known that the Russians were indeed advancing, Napoleon wrote to one of his ministers, 'All my intelligence told me that the Russians retreated when they learned what had happened to Prussia.' It is to be noticed that when he became certain that the Russians were coming to meet him he considered this whole treaty as inadmissible and immediately cancelled all its arrangements.

The King of Prussia found these terms much more exacting than he had anticipated. He had thought of a speedy forgiveness and then reconciliation, and a return to the fold. On the 26 October he sent Count Zastrow to Charlottenburg to join Lucchesini in pursuing the negotiation and to make a strong attempt to secure better terms of peace. He 'could not disguise' from the Emperor that the severity of his stipulations had distressed him all the more in that the letter he had received after Jena had given him hopes of a reconciliation on more easy terms. He talked again of his 'sincere desire to re-establish the old relations with France'.[10]

Napoleon gave Zastrow a friendly interview and spoke kindly of Haugwitz. On the 28th of October the two Prussian negotiators recommenced the discussions with Duroc. Two days later, in a note to that minister, they accepted the bases which Napoleon dictated for a peace – bases in some ways modified from those given at Wittenberg a week before.[11] The French demands had not become more merciful or kind; Napoleon gave it to be understood that he would require Frederick William to forbid the entry of Russian troops into his territory and to engage to fight at the side of France in the event of a war with Russia; the Prussian king felt the conditions grievously hard and was very unwilling to accept them. On the 6 November a conference was held at Graudenz to discuss the advisability of submitting; the whole assembly was of the opinion that it was out of the question to think of renewing the war, and Haugwitz put forward the idea that the slightest hesitation or delay would provide Napoleon with an excuse for raising his demands.[12] Everything was done to prevent or to evade the necessity of having to fight Russia, whose armies were by this time unmistakably advancing into Poland. It was decided to attempt to persuade the Czar to avoid hostilities, and a messenger was sent to St Petersburg to make the treaty palatable to the Russians. On the 7 November Frederick William wrote a letter to Napoleon:

> In spite of the dreadful sacrifices which you, Sire, have just imposed upon me, I do not desire any less sincerely that this peace, already assured by my acceptance of its bases, shall enable me soon to reestablish with Your Majesty the friendly relations which a moment of warfare has suspended.... I have not even waited for the signature of the treaty before stopping the advance of the Russian troops.[13]

These Prussians, it seems, were anxiously holding their breath, lest Napoleon should change his mind again and revoke the agreement he had offered. Haugwitz, groveling to the victor, still wrote to Lucchesini to induce him to seek an amelioration in the terms of the treaty; Prussia must not be made too weak to act as a check against Russia, he said; Prussia was only too anxious to be at the service of France; and he pressed the idea and fawned like a servile courtier, engaging his word 'as a man of honour, who for twelve years has consistently worked to cement the relations between France and Prussia.' If only Talleyrand would appear, thought Haugwitz – he at least could work on lines less wild and impolitic 'than that terrible principle of the destruction of Prussia to guarantee the future repose of France'.[14] Talleyrand distrusted some of the drastic measures of his master, and loved to tone them down, and regarded the victory of Jena in the moral effect it had upon Napoleon as tantamount to a disaster for France. Lucchesini also thought that in him lay the one hope of Prussia. Prince Ferdinand, great-uncle of Frederick William, sought an interview with the conqueror, wrote 'as the brother of Frederick the Great, as the oldest member of the family, and as the oldest soldier in Prussia', in order to beg for pity–wrote also to Murat to ask for the amelioration of the peace terms; but his efforts were ignored. There was some talk even of Frederick William himself going to make personal appeal to Napoleon. It was in vain to try to bend or soften that rough imperious will – nothing was to be done save to reconcile oneself to the great sacrifices demanded and hope that now Napoleon would at least not change his mind and harden his heart anew – would not retract his offers and hurl the whole question into the melting-pot once more.

Having agreed to the treaty the Prussians still trembled, and for some time they were kept in suspense. The actions of Napoleon became strange and inconsistent; the shadow of the man loomed threateningly again; and every sign in the sky seemed malevolent. The negotiators at Charlottenburg were left to themselves. Their note of the 80th of October, accepting the terms of the French treaty, dropped into an awful void and never had an answer. The Prussians found themselves helpless and bewildered before one of those long inauspicious silences of Napoleon – silences of a man who was brooding new terrors for the world. The Emperor had decided to change his plans.

* * *

The whole history of the year 1806 proves that the real antagonism in Europe at this time had been between France and Russia. Throughout the year both powers had turned their attention to the Adriatic and the Near East and had seen there the next, the most immediate, conflict of their interests. The Czar had watched anxiously the eastward extension of Napoleon's power in the Mediterranean and was jealous of the growing influence and supervision of France over the politics of Constantinople; Napoleon could hope, in fact, that he would declare war against the Turks in order to destroy the French predominance there. For some time Napoleon, when he had concluded treaties with various powers, had made a cardinal point of securing a clause, apparently innocent but really directed against Russia, guaranteeing the 'independence and integrity' of the Ottoman Porte; that is to say, he envisaged a war with the Russians that should be fought ostensibly on behalf of Turkey. So long as there was a power like Prussia, even half-unwillingly and merely ostensibly allied to France, but acting as a barrier on the Polish frontier of Russia, the government of St Petersburg would find in it one reason for carrying on the conflict with France not in that region but in the Turkish Empire. Prussia indeed by being a buttress in the east had had her place in the Napoleonic system in Europe and had formed part of that disposing of forces which Napoleon had hoped to maintain. That was why he had considered war with her as a 'civil war' and could not forgive her for forcing it upon him.

From the moment when the problem of Prussia had become serious, the Russians had never ceased to occupy Napoleon's attention. They had denounced a peace treaty signed by their plenipotentiary in July, and it was plain that they would find in the new crisis their own opportunity. Over and over again Napoleon had betrayed his apprehensions. He seems to have been very anxious that the Russians should not come to meet him in Poland. It would give him war in the wrong quarter: 'My concern with Russia is in connection with the Porte',[15] he had written to Talleyrand in July. Moreover, it meant a winter campaign in a region which by its climate and geography made warfare difficult for Frenchmen and victory only less miserable than defeat. Circumstances were particularly unfavourable for such a struggle if France had to fight without the intended assistance of Prussia. Perhaps, finally, it is not fanciful to say that Napoleon saw himself deprived of what he had conceived to be a chance of gaining a foothold in the Ottoman Empire. In any case he made it clear to the Prussians that their fate was to depend upon the conduct of the Czar.

A proclamation issued to the French army on the 26 October contained these words: 'The Russians boast that they are coming upon us; we will march to meet them; we will spare them half the journey. They will find Austerlitz this time in the middle of Prussia.' On the 31 October the advent of this new enemy is still rather doubtful. 'We do not hear any more talk of the Russians,' the bulletin runs; but its challenging tone shows that there exists an idea that they will come into Poland. 'We are longing for a hundred thousand of them to come. But the rumour of their advance is mere nonsense. They will not dare to come to meet us.' Four days later, however, the question is decided: 'The Russians are a long way off; but it is possible that we shall meet them and be at grips with them in a month's time.' This is the moment that finds Napoleon breathless, and the whole man flashes to life. On this day, the 3 November, he writes urgently to Paris for reinforcements. He sends for Kosciusko, the Polish rebel, to incite his fellow-countrymen against the Russians and henceforward his correspondence is full of the idea of an insurrection in Poland. At this time Talleyrand whips round to the court of Vienna, where armaments have been made to guard the territory and enforce the neutrality of Austria. These armaments have suddenly taken on an appearance of menace, so Talleyrand writes to the French minister there, writes directly to the Austrian chancellor himself,[16] to demand their instant cessation. Altogether these first days of November are a pivot. Once more Napoleon shows himself alert, breaks out with unbaffled energy, and nerves himself for a crisis. A certain situation has arisen and all the schemes and expedients that were to have been contingent upon it are now given release. Here is something that makes the peace-terms of November inadequate. There is to be a war with Russia in Poland after all – and this war produces a new situation in Europe and leads to a drastic reconsideration of many of Napoleon's plans.

On the 25 October the French entered the Prussian capital, and at this time they in Berlin and Napoleon in Charlottenburg discovered documents and correspondence which revealed the solidarity of the Prussian connection with Russia and linked up the two nations more closely than Napoleon had even suspected, and entirely dissipated what doubt there had been concerning the existence of an understanding between the two courts. 'Napoleon has found a letter in Charlottenburg,' wrote the Countess Voss, 'left by the Queen, or, as they say, slipped behind a sofa cushion where it had remained unnoticed, and this letter has made him quite furious.' These compromising documents convinced him more than ever of the treachery as well as the unwisdom of the conduct of the court of Berlin in the summer of 1806. In his bulletin of the 26 October he begins to put his finger upon the connection between the Czar Alexander and the Prussian court as the cause of the war policy of Prussia and he no longer repeats former statements about an unwilling king being rushed into hostilities by a war party in his cabinet. From this moment he bursts out in bitterness against the queen of Prussia and advertises her friendship for the Czar and attacks her with gratuitous insults in his proclamations. If it had before seemed a serious matter to fight the Russians in Poland with a defeated and embittered Prussia sulking in the rear, the objections to this course were redoubled now that it had come to light that Prussia had for a long time been pursuing a dual foreign policy, balancing her alliance with France by a more secret and intimate connection with the Czar.

The prospect of a campaign in Poland had raised a further issue that was likely to make Napoleon withdraw the offer of peace that had been made to Prussia. The whole problem of 'the barrier of the east', the buttress against Russia, had been in Napoleon's mind ever since he had come to issue with the court of Berlin, and the approach of the Russian armies in November brought the question to a further stage, since it put Poland more directly under the eye of Napoleon and gave him a more immediate interest in the idea of reviving the kingdom. The national feeling of the Polish people would be at his service in a war against Prussia and Russia, if he could convince them that in return for their assistance he would undo the work of the partition treaties and restore their state to its former independence. It would be the revival of the old connection between France and Poland, and much that was sentimental in the relations between the two nations tended to make such a policy attractive. France had been helpless to prevent the destruction of Poland, but now that Napoleon was striding Europe as an arbiter, remodelling the continent and doing feverish mapmaking, it was natural that there should be a certain expectancy in the air, and a feeling that France would do her part at last.[17] Opportunist as he was, Napoleon certainly encouraged the Poles to hope for salvation in this way; he dangled the scheme for the restoration of Poland before their eyes; it suited him to take the point of view that 'France has never recognised the different partitions of Poland'; and, without promising anything, he hoped at least to secure Polish assistance in the forthcoming campaign against Russia, by merely keeping alive the vague expectations. Further than this he seriously considered the possibility of actually adopting the policy that the Poles desired and reviving the Kingdom of Poland. He saw that this revived kingdom might be made a client state of France as in the 18th century, and thus might be used as a substitute for the Prussian barrier in the east. The idea was one which would demand a readjustment of the peace terms he had already offered to Prussia; but, further, it was a course which in any case rendered the future alliance of Prussia less important to him, and made immediate peace impossible. The emergence of the Polish question necessarily threw the fate of Prussia into the melting-pot again. It was a further reason for cancelling the arrangements already made.

The restoration of Poland, then, became an open question, but Napoleon did not definitely commit himself to any scheme as yet;[18] the subject was one upon which his own entourage seem to have had no unanimous opinion. The prolific writer Montgaillard presented on the 5 November a project which comprised the re-establishment of Poland, 'if it is no longer allowed to be hoped that the cabinet of Berlin will oppose a serious resistance to Russia'. Montgaillard thought that after all the humiliations it had suffered the Prussian royal family would never be reconciled to imperial France.[19] Jomini, however, a few days later was urging the merits of an alliance with Prussia, and showing what might be won from the gratitude and friendship of Frederick William if the Prussian lands could be increased by as much of Polish territory as could be scraped together for them.[20] Napoleon at St Helena made claim to have considered such a scheme, declaring that he had desired Hanover, the Prussian provinces of Saxony, Westphalia, and Franconia, and Magdeburg, for himself; but asserting that he would have made Frederick William the ruler of a restored Poland, and would have given part of the Illyrian provinces to Austria in return for the cession of Galicia.[21] It is more than likely that he played with such schemes for the future of Prussia – halting at them, perhaps, for a moment, as his mind went through the kaleidoscope of the possibilities before him, and then idealising them in his memory during exile. But what seems irresistible as the solution of the Polish problem in the later months of 1806 is that Napoleon still held everything fluid, and preserved the mobility and uncommitted freedom of his mind, and was prepared to do anything as he might be called upon to adjust his policy to the exigencies of a growing situation.

At any rate he would have nothing more to do with the peace treaty he had offered to the Prussians. All that had happened since he had proposed the terms of it, all that he had learned of the duplicity of the court of Berlin, impelled him to make a different and a more drastic use of his recent victories. This is where the astonishing efficiency of his subjugation of Prussia came into use. After Jena he had remorselessly pursued the remnants of the Prussian armies and had captured the Prussian strongholds. Many victories had been given to him by the shameful conduct, both the cowardice and the treachery, of his enemies. Towards the end of October and in the first days of November news of the capture of Hohenlohe's army, of the capture of Stettin and Cüstrin, and of the fall of Magdeburg and Lübeck reached Napoleon, so that on the 16 November he could announce: 'the campaign against Prussia is entirely finished'. Duroc stated that if only the news of the fall of Magdeburg had not come through at an inauspicious moment peace might have been concluded. But Napoleon, having astonished himself by the ease and the completeness of his victory, determined to exploit his success and make more effective use of his advantage than the terms he had already granted would allow. The arrival of the Russians in Poland and the amazing facility of his recent conquests had produced a great extension of the scope of his plans.

On the 9 November Napoleon decided to recommence the negotiation that had been suspended, and ordered Duroc to meet Lucchesini and Zastrow at Charlottenburg once more in order to tell them

> that I consider the battle of Jena as terminated, now that the corps which had been cut off at Lübeck have surrendered, and that in this state of things I do not see anything to prevent us negotiating seriously for the conclusion of peace. The necessary preliminary seems to be an armistice such as they have requested from M. Talleyrand.[22]

The former negotiations and preliminaries and outlines of peace now counted for nothing and were swept away as an outworn pastime; henceforward the two powers were to settle down to 'negotiating seriously.'

On the 16 November the Prussian plenipotentiaries, Zastrow and Lucchesini, signed an armistice which handed over to Napoleon all that his armies had hitherto failed to secure. The Prussians engaged to withdraw their troops to Royal Prussia and the region around Königsberg, and to surrender part of the province of Silesia together with a large number of fortresses that the French had still failed to capture. It was also stipulated that parts of East Prussia should be occupied by neither Prussian nor French troops. A clause of the convention announced the continuance of negotiations for peace at Charlottenburg, and provided that, in case of the failure of these, hostilities should not be resumed by either party without a warning of ten days being given. In addition to these things the King of Prussia was to expel any Russian troops that had entered his territory and to engage not to receive them on Prussian soil during the armistice; that is to say, he was to commit hostilities against the very ally he had called to his aid.[23]

Napoleon described the suspension of arms as 'only a military measure, which serves to arrange the question of winter in a season so advanced'. At the same time he ordered his marshals, Davout and Lannes, to make no announcement of its signature until the fortresses were actually given up to the French.[24] He was anxious concerning the Poles whose aid he was invoking against the Prussians and Russians; he knew they were jealous of his negotiations with Zastrow and Lucchesini. They were afraid lest, having compromised them; he should desert them and make peace with their oppressors. If he could delay informing them of the armistice until the actual execution of the terms, then, he said, 'it will be easy to make the Poles see that by the suspension of arms, their existence is partly recognised.... In the long run everything is favourable to their cause', Talleyrand wrote:

> The conditions (of the armistice) are such that Poland, if there is a Poland, will recover the liberty of having and expressing an opinion. Our troops will not withdraw from the part of Polish Prussia that we already occupy; and it is agreed that there will be neither French nor Prussian troops in the other parts of Polish Prussia that we do not hold. So you have Polish Prussia considered as a separate thing from Prussia; and, moreover, you have a Poland that can show itself, if it has the means... Poland which menaces Russia, and by this can lead her to the peace we desire, that is to say a peace that preserves the Ottoman Porte.[25]

From all this it appears that Napoleon has by no means committed himself to a restoration of Prussia in any future settlement of Europe. If he has not committed himself to the Poles, either, he still leaves himself room to play with them, and opportunity for stirring their hopes. He has merely reaffirmed his desire to have, on the frontiers of Russia, a power that will preserve and uphold his interests in the Ottoman Porte. One wonders what he could have expected the Prussians to see in the armistice to make them welcome it as safety and reassurance and hope. One wonders why the negotiators dared to sign such a document at all. Lucchesini defends himself by saying that they were held like prisoners at Charlottenburg, that their correspondence with the Prussian court was uncertain, that they could not obtain correct information of the true situation of things. They thought the Russians were still far away, and, in that case, they knew that the king could ratify the armistice, and when the Russians arrived find an early pretext for breaking it; but even if the Russians were approaching they felt that the king could refuse to confirm their work. They were afraid of an insurrection of the Poles, and they counted that a suspension of arms would discourage this. They were distressed that the estates of the Prussians seemed condemned to be the theatre of the new war. The Czar Alexander, it was thought, would be ready for peace if Napoleon decided not to attempt the restoration of Poland. Talleyrand begged the Prussian negotiators to make the Czar see the dangers that would fall upon Prussia, if Russia persisted in the war – the danger there would be of Russia herself being thrown beyond the Dwina if the risen Poland and the provinces conquered by France should receive a chief who would be a mere vassal of France. It was further insinuated to Zastrow and Lucchesini that it was very important for the restoration of Prussia to make Napoleon commit himself to something by the signature of a document such as this. Under such arguments and apprehensions the Prussian negotiators signed the armistice; and, wrapped up in the same insinuations and insidious suggestions, the document came to the King of Prussia, inviting acceptance and ratification.[26]

In a note which came as an appendix to the armistice the real schemes and intentions of Napoleon were made apparent; and it was revealed that the armistice itself was only incidental to the new development of Napoleonic policy. Lucchesini afterwards related that he knew nothing of this note until the suspension of arms was settled and signed; otherwise, he maintained, he would never have put signature to the agreement. Duroc, it seems, received the note suddenly by messenger from Talleyrand, and handed it to Zastrow, who read it and seemed pleased with it, evincing what many contemporaries would have thought a typically Prussian reaction to its contents. 'So much the better,' he said, 'this gives us some more means of forcing the Russians to enter into negotiations with France. Anyhow the treaty is signed.' The note recalled the generous conduct Napoleon had pursued in giving up conquests he had made after the collapse of three successive coalitions against him.

'But in the course of wars perpetually renewed, France, Spain and Holland have lost their colonies. It is natural, it is just, that the countries which the right of conquest has put into the hands of the Emperor, should serve as compensations for the Colonies.' In a similar way 'His Majesty cannot give up any of the conquests which the chances of war have put into his power before the Ottoman Porte is completely restored to all its rights over Moldavia and Wallachia, and before its absolute independence is recognised and guaranteed.' In other words, Napoleon refused to conclude separate peace with Prussia.

In his message to the Senate a few days later he outlined the same policy with similar logic – talked again of his extreme moderation in previous wars, and remarked that 'a large number of the Cabinets of Europe are sooner or later influenced by England, and without a solid peace with this power our people will not be able to enjoy the blessings which are the first aim of our labours.' He stated that he would not evacuate Poland and Berlin 'before a general peace was concluded, the conquered colonies restored and the independence of Turkey guaranteed', and he made it clear that he was ready to treat for peace with England, Russia and Prussia 'but only on bases such as do not permit anybody to arrogate to himself any rights of superiority over us, and such as give back the colonies to their Mother-Countries, and guarantee our commerce and industry the prosperity they have a right to'. This same message to the Senate announced also the creation of 'the continental system', which was intended to place Great Britain in a state of blockade and to put her under the ban of Europe.

Napoleon then acknowledged that England, out of reach of the French armies, could be a perpetual foe of France and the source of repeated wars in Europe; but he met the problem by the establishment of a new and sweeping principle of policy. The idea rested on the carrying out of a series of thorough but conditional and temporary conquests in Europe, which were to be the pledge of a general peace. Napoleon would not subdue Prussia for ever or annex the state finally into his Empire, but he would occupy the territory until his enemies should find it to their own advantage to redeem it. To end the French occupation the Prussians would set everything in motion for the cessation of the war. The Czar had been willing to come forward to fight for the salvation of Prussia; now he would be called upon to make peace for the identical object. The continent would find the sufferings of the war so heavy that Europe would range itself against England on behalf of tranquillity. To restore the Prussian monarchy, raise up a barrier against France and bring relief to Europe, the Russians would abandon their thoughts of interference in Turkey, and the British would surrender their colonial conquests. Everything was comprehended in one trenchant principle of policy, for Napoleon had a way of fusing his problems together and resolving them at a stroke, covering the whole political situation by an imaginative leap, a bold synthesis.

It was all a new mapping out of the European situation, a fresh disposing of the forces at play, a great European design that fits in with the other 'continental system' which was decreed a few days later at Berlin. Napoleon's attempt to boycott English trade and shipping represented the same kind of conspiracy to range the continental powers against England. Europe was to suffer for England's persistence in the war, and was to be the accomplice of France in bringing the British government to heel. Such plans resulted in a heavy increase of Napoleon's obligations and burdens in Europe; for they demanded that a large extent of territory or coastline should be under the occupation or control of France. They had in fact to be extended before they could come into effective operation against England; and until the alliance of Russia was secured their working was not sufficiently drastic. But there is a sense in which these plans of November 1806 are a half-way house on the way to Tilsit. They represent a stage in the development of the Napoleonic design of beating England through a control of the continent.

In this new organisation of policy the Mediterranean does not occupy the same prominent place as before; the ambitions that centred in Turkey and the east are not in the forefront for the moment; Napoleon has had to change his plans. He no longer holds the picture of a Russia deterred on the side of Poland by the presence of a Prussian barrier and coming to grips with France in the south of Europe. He has a different map in his mind. He accepts the war in Poland, accepts the defection of Prussia – will not even allow this power to return to his system. But he changes his old difficulty into a source of power, for Prussia in the ranks of the coalition is still to serve his purposes and promote his views, and will do so more effectively than if he permitted her once again to become an embittered insincere ally. Prussia is to be a weakness and a burden to her new friends, a complaining partner in the war and a dejected advocate of peace. She is to be this because he has given her the motive and has provided her with just the appropriate cue. Once he has made it known that the Prussian king shall not return to Berlin until Russia and England have come to terms, he can leave the situation to work itself out. The rest is automatic.

Napoleon, having achieved the conquest of this state, had no intention of retaining the country as a permanent part of his acquisitions. He had been at a loss at first as to the use he might make of his good fortune – had even talked of deposing the house of Brandenburg and enthroning another reigning family in its stead. For the future, however, he determined to make a military advantage out of his occupation of the territory; he needed it in his forthcoming campaign in Poland. Further than this, he had found a way of using the conquest as a diplomatic weapon, as a thing of negotiable value, in his dealings with the powers. So Prussia was not allowed to make peace, was not allowed even to accept defeat; she was thrown back into the coalition which she was seeking to escape.

This statement of policy came as an appendix to the armistice that Zastrow and Lucchesini signed on the 16 November 1806. That armistice had still to be ratified by the Prussian king. According to its sixth article the ratifications were to be exchanged by the 21 November. Duroc himself was sent to the Prussian king to secure the acceptance of the arrangements. His instructions were peremptory; if the slightest objection were made to the terms he had to propose he was instantly to return. On the 21 November the whole question of the armistice was discussed by the Prussians in a conference held at Osterode. The military men present expressed the opinion that the stipulated surrender of the fortresses would not be a serious blow to Prussia; as such a loss would be inevitable in the chances of continued war; while the suspension of arms would give the king an opportunity to raise troops. Stein, however, who had been in communication with Hardenberg, pointed out the discrepancy that existed between the terms of the armistice and the general policy of peace that was expounded in Talleyrand's note of the 16th; he showed that if the future of Prussia depended upon the willingness of England and Russia to make sacrifices for her, it could not depend on the conclusion of an armistice that would only alienate these powers from the Prussian cause. The armistice and the note that came as an appendix to the armistice were not complementary but contradictory statements. This was the most disastrous side of the convention which Zastrow and Lucchesini had signed. At the very moment when Napoleon finally and decisively set down Prussia as belonging to the coalition, making her a factor in the system of compensations that England and Russia were to accept at a future peace and so binding up her fate irrevocably with theirs, he asked the Prussian king to engage to commit hostilities in favour of France. The armistice that was offered as a promise of peace contained really a new commitment to war. It was all part of Napoleon's way of giving with one hand and taking away with the other. Stein riddled this fallacy of the armistice. He laid bare another piece of Napoleonic trickery, showing that it was not in the power of Prussia to fulfill the terms of the convention and expel the Russian troops, and that if the armistice were signed Napoleon would soon discover this and denounce the agreement; by which time the fortresses would have been handed over to him and so lost beyond recall. To this reasoning the cabinet councillor Beyme contributed further by showing how the engagement that no troops of either power should remain in East Prussia was one which also told in favour of France, since that region would be affected by the Polish insurrection and so would be lost to Prussia by the armistice. These arguments eventually carried the day.[27] The armistice was rejected. The king placed his fate and future in the hands of Russia. He did not risk much by this, 'for the armistice would have robbed him of nearly everything'.

Within a few days Duroc brought to Napoleon the news that Prussia, finding the Russian troops already within the borders, had had no choice but to reject an armistice, the terms of which she could not hope to carry out; and a little later Zastrow and Lucchesini received the orders of their court. Lucchesini was commanded to remain with the French and to continue the negotiation for peace in any locality that might be agreeable to the French government. For when Prussia rejected the armistice she did not intend by this to imply that she plunged wholeheartedly into the war against France. She took the step because she had decided that if she had to fight it must not be against Russia. She refused to ratify not because she repudiated or abandoned the idea of ending the war but because she discovered the trickery of the armistice and found that it was not a step towards her object. The two Prussian plenipotentiaries were directed by Talleyrand[28] to Posen, where the Emperor had just arrived. On the 6th of December they had their interview with him.[29] Napoleon declared to them that since the Prussian king had rejected the armistice which they had signed and had disavowed their work it was obvious that they had not the confidence of their court, so he would negotiate with them no longer. He stated that if his armies met the Russian forces in Poland military reasons would not allow him to restore any part of the dominions of the king of Prussia. There could be no negotiation with the Prussians; for, desiring the recovery of their territories, they had nothing with which to negotiate for them. The only reason which could induce France to listen to their overtures would be the hope that they might draw into the transaction other powers like Russia and England with whom there might be made some bargain or exchange. More clearly and decisively than on the 16 November it was shown to the Prussian plenipotentiaries what Napoleon's new policy implied for their government. The future of their nation was to depend on the success of their persuasions with other courts on behalf of a general peace.

Thus, in the beginning of December, Prussia is a member of the coalition against her will, and finds herself at war with Napoleon in spite of herself. Events have reached their full cycle and she has discovered that her action in declaring war against France has produced a situation which prevents her from turning back when she desires. The very coalition which she has called to her aid and for which her conduct is at any rate the signal and the opportunity, has become – as Napoleon foresaw – the very thing which forbids her resumption of former relations with France; and she finds herself more closely and irrevocably bound up with its fortunes than she had ever dreamed of becoming. For a long time she had refused Napoleon's offers of separate arrangement with her, she had insisted on having her war, until there had come a time when Napoleon, face to face with Russia, could not safely admit of a separate treaty with her. It was not merely that he was proud and angry and intoxicated by easy victories – though these things perhaps encouraged him to plans that were too large and to enterprises that were to be too great a burden to France; and such personal factors have weight in history as much as the logic of policy. But Prussia had produced a fresh situation and a fresh problem for Napoleon. For it was not a case of simple arithmetic, of plus and minus, of a coalition of three turned into a coalition of four; the entry of Prussia into the ranks of Napoleon's enemies had reversed a system and smashed a synthesis, and upset the chart of forces that Napoleon had mapped out for Europe. And the new peace policy outlined by Talleyrand on the 16 November, coming almost simultaneously with Napoleon's real decision to isolate England by measures such as he had toyed with and had revolved in his mind and had even effected in an experimental way before he plunged into them with the decree of Berlin, was really a new synthesis, a fresh way of envisaging the problem and of mapping out the relations of Europe, that Napoleon worked out for the changed circumstances.

It was observed at Osterode when the king rejected the armistice that he did it 'with regret, not with indignation.' Prussia had begun the war of the fourth coalition; Napoleon was compelling her to see it through. But she was an unwilling member of the coalition, scarcely more than a mere passenger in it. That was her disposition when December came in, and the campaign in Poland began.

Notes

1. *Lettres et documents pour servir à l'histoire de Joachim Murat*, iv (1910), 391.
2. *Private and confidential... narrative of the circumstances which led to an interview between the Emperor Napoleon and G. Sinclair*, by Sir George Sinclair (1826).
3. *Correspondance de Napoléon 1st*, t. XIII, No. 11,031.
4. L. von Ranke, *Denkwürdigkeiten des Staatskanzlers Fürsten von Hardenberg*, III (1877), 211; 16.
5. Lucchesini's account of his mission in *Bailleu, Preussen und Frankreich*, 1795, bis 1807, II (*Publikationen aus den K. Preussischen Staatsarchiven*, Bd. XXXIX), 631–4.
6. *Corresp. de Nap. 1st*, t. XIII. No. 11,045.
7. Rapp, *Mémoires écrits par lui-même* (1823), p. 118.
8. The letter is quoted from the French Archives by Driault, *Tilsit* (1917), p. 41, and (rather differently) by Lefèbvre, *Histoire des Cabinets de l'Europe*... (1866), II, 399–400; so presumably it was dispatched; but cf. Ranke, *Hardenberg*, III, 219 (note).
9. For this treaty see Bailleu, *Preussen und Frankreich*, II, 577.
10. Letter of Frederick William to Napoleon, 26 Oct. 1806, in Ranke, *Hardenberg*, III, 219–20.
11. Ibid., III, 221.
12. Ranke, *Hardenberg*, v, 396, Protokoll der Konferenz abgehalten in Graudenz am 6 Nov. 1806.
13. E. von Höpfner, *Der Krieg von 1806 und 1807* (1855), Part I, vol. II, p. 389; A. Lefèbvre, *Histoire des Cabinets de l'Europe 1800–1815*, II (1866), 402–3.
14. Haugwitz to Lucchesini, 7 Nov. 1806, in Bailleu, *Preussen und Fra nkreich*, II, 578–9; cf. Hauterive to Talleyrand, 19 Oct. 1806, ibid. II, 612-13: 'Je vois, mon cher ami, que vous n'êtes pas persuadé encore que la Prusse est une puissance vile.'
15. *Corresp. de Nap. 1er*, t. XII, No. 10,448.
16. Talleyrand to Stadion, 2 Nov. 1806, A[rchives du ministère des] A[ffaires] É[trangères], Autriche, t. 379.
17. *Corresp. de Nap. 1er*, t. XIII, No. 11,279.
18. Ibid., t. XIII, Nos, 11,258, 11,279; t. xv, No. 12,603.

19. Clément de Lacroix, *Mémoires diplomatiques de Montgaillard 1805–19* (1896), pp. 307–21.
20. F. Lecomte, *Le Général Jomini; sa vie et ses écrits* (1860), pp. 51–4. Jomini, *Vie Politique et militaire de Napoléon racontée par lui-méme*, II (1827); 328–9. See also proposals of Hauterive in Bailleu, *Preussen und Frankreich*, II, 581–2 (footnote); II, 612–13; Driault, *Tilsit*, pp. 68–9. On the whole subject of Poland see Handelsman, *Napoléon et la Pologne 1806–7*, pp. 1–22 and 167–76.
21. Corresp. de Nap. 1er, t. XXXII, p. 443 in the quarto edition (p. 358, octavo): 'Extraits des récits de la Captivité; Tentatives pour le rétablissement de la Pologne,' 20 Aug. 1820; cf. however, 'Entwurf einer Proklamation Napoleon's über die Absetzung des Hauses Brandenburg (Ende 1806),' Bailleu, *Preussen und Frankreich*, II, 581–3.
22. Instruction for Duroc, 9 Nov. 1806, in Bailleu, *Preussen und Frankreich*, II, 579–80.
23. *Corresp. de Nap. 1er*, t. XIII, No. 11,277.
24. Handelsman, *Napoléon et la Pologne 1806–7*, p. 21.
25. Handelsman, *Napoléon et la Pologne 1806–7*, pp. 20–1. See also Talleyrand to Andréossy, 18 Nov. 1807, printed ibid., p. 215.
26. A précis of the apology which Lucchesini made for the signing of the armistice can be found in Ranke, *Hardenberg*, v. 421. See also F. v. Ompteda, *Nachlass* (1869), p. 248 (No. 152): 'Notatum des Gesandten Ompteda über eine vertrauliche Unterredung mit dem Marquis Lucchesini (Wien) Ende Jan. 1807.' It will be noticed that Lucchesini had found his way to Vienna, where, according to Razoumovski, his conversation was hostile to Russia. A. Wassiltchikow, *Les Razoumovski*, t. II, 3me partie (1894), ch. III, pp. 81–2.
27. Ranke, *Hardenberg*, v, 398. See also *F.O.* 64/71, George Jackson's reports Nos. 10 and 11, 21 and 23 Nov. 1806. G. Jackson was the semi-official British agent with the Prussian court, England being at that time still technically in a state of war with Prussia. His correspondence (*F.O.* 64/71) and his diaries (II, 29–54) throw some light on this period.
28. Talleyrand describes his interview with the Prussian negotiators on the 29 November in P. Bertrand, *Lettres inédites de Talleyrand à Napoléon* (1889), No. CCXVII.
29. For this interview see G. Jackson to Howick, No. 17, 16 Dec. 1806, *F.O.* 64/71; cf. *Corresp. de Nap. 1er*, t. XIV, No. 11,394.

4

Austrian Policy and the Austrian Attempt To Bring About Peace

Andréossy, French minister at Vienna, was a military man and a bitter anti-Austrian – far more bitter than Napoleon who at this time seems to have paid little attention to his outbursts. He took a cruel pleasure in recording and perhaps exaggerating the anguish of the Austrians in the critical weeks of July and August. 'The news of the battle of Friedland made very little impression on the Court and Ministry', he wrote. 'It was not so with the news of the armistice...This has caused a most disagreeable sensation and inspires the most lively fears.'[1] Austria, he reported, was continuing her armaments; she was full of fear and loneliness, and felt that her hour had come. He gloated over her discomfiture. Concerning the intentions of France he kept up a deliberate air of mystery, hoping thereby, as one Englishman said, 'to draw the Austrian cabinet into some decisive measure which might furnish Bonaparte with an excuse for turning his arms against them'.[2] Andréossy's despatches were written in the following strain:

> The silence which is kept towards her is giving her anxiety, and she feels humiliated that she, mediating power, was not warned in any way.... It will be possible, I think, to make any demand of Austria that may suit the arrangements of His Majesty. The more we adopt a high-handed tone befitting the prodigious ascendancy which His Majesty has acquired, the more easily we shall obtain demands which in other circumstances would have been difficult to obtain or would have been passed over in silence... [Austria] is afraid of her isolation. they seem to wish for a rapprochement with France and are disposed to be condescending.[3]

A glance at the Austrian minister in St Petersburg would confirm this impression of the nervousness of the court of Vienna. Merfeldt also was a military man, an unfortunate diplomat to employ. While the treaties of Tilsit were being arranged he had heard rumours that Russia was to have increase of territory at the expense of Prussia and Austria. He had asked the acting foreign minister at St Petersburg to inform him of the tenour of the treaties; this in return for the 'unlimited confidence' which the court of Vienna had always given to the Czar. Merfeldt conjectured that Napoleon would have a grievance against his country, which had so often declined a French alliance; he insinuated that, 'Whatever desire the House of Austria might have to remain at peace with all its neighbours, it was far from shrinking from the risks of a new war. The Austrian monarchy had perhaps never been in so flourishing a state as at the present moment.'[4] Tactlessly at a later date Merfeldt addressed himself to the French minister Savary, and spoke with some indiscretion. According to Savary he was 'furious' that peace had been made. He made wearisome demands to learn the terms of the arrangements. He said, 'I confess that it is difficult for me to believe that you can have made a lasting peace without us'; and when Savary replied that he did not know of any questions that remained to be settled with Austria, he declared, 'We cannot be indifferent to the fate of Wallachia and Moldavia', and even made some wild references to war. 'Allow me to say', replied the French minister, 'that your imagination sometimes makes phantoms for itself in order to have the pleasure of fighting them. At this moment you are already seeking a motive for war in a treaty that has hardly been signed.' The Austrian seemed discontented with the settlement of the Polish campaign at Tilsit. Savary referred to the neutrality of Austria. 'Had you an indirect part in this War? Were you interested in seeing it prolonged?'[5] Merfeldt, who had acted unwisely at a time when it had been important to promote good feeling between Austria and Russia, was now restless and over-anxious, and too good a coalition man. He made himself too ostentatiously a member of the English party at St Petersburg – showed more respect to the British ambassador than to the minister of Napoleon. 'If the English ceased to have a legation here', wrote Savary, 'M. de Merfeldt would watch over their interests'. So Napoleon demanded Merfeldt's recall, and the Austrian chancellor had to complain that 'Russia treats us with coldness, or rather with reserve.'[6]

Starhemberg in London was also an enemy of France and of the Tilsit system. In the early days of 1807 he had been severely reproved for his excess of zeal in confiding to his official despatches to Vienna sentiments that were of a nature to compromise his government with France. He, also, was nervous for his country after the battle of Friedland, and, stepping beyond his instructions, he had 'several friendly but not official conversations with the Secretary of State on the possibility of our being attacked at some time without provocation'. Canning took the question seriously, and the result was a situation of some irony; Roles were reversed for the moment, and it was the British minister who urged the Austrians to keep the peace at all costs and avoid any differences with France. Austria could not be considered as having the same incontestable right to the assistance of England as if she had come to the help of Russia. 'Only in the case of a war of self-defence in the strictest meaning of the words' would England give any consideration to the idea of helping the court of Vienna. Canning deplored that the Austrians had distrusted Russia more than they had feared Napoleon; he urged them to make a close union with the Czar; this was the system he desired to see established in Europe; the king, he said, recommended this policy, 'at a moment when, if the rumours of the Continent are to be credited, he has himself to apprehend at least an unfriendly disposition, if not a hostile determination against him, on the part of Russia'.[7]

The government at Vienna had felt its fate hang in the balance when the conclusion of the treaties of Tilsit had left Napoleon free to deal with Austria. For a time there had been a breathless suspense, as men waited to learn what Napoleon's decision might be. It was no question, for the moment, of coalitions and conspiracies against France, of measures to check the usurper, of wide ulterior schemes; the Austrians held their minds feverishly centred on thoughts of self-preservation. The duplicity of two years had seemed about to receive its answering stroke; and if anything could be judged from the tone of the French minister, Andréossy, the stroke threatened to be vindictive. Then the silence in which Andréossy had seemed to exult, the cruel suspense, had broken. Napoleon declared to General Vincent that he was pleased with the late conduct of Austria; Duroc wrote to Andréossy in the same friendly spirit; and Talleyrand reported that the Emperor was disposed to enter into explanations with the court of Vienna on all matters that required to be settled.[8] This meant the end of a crisis, but not the end of nervousness, for Austria. She could not immediately readjust her psychology and leap out of the tense atmosphere and lose her cowering dread. Everything that had happened had only confirmed this power in a policy based upon fear. She did not need a Canning to warn her against offending France; the Austrian chancellor quickly retorted to such advice that it was his intention 'not to give France any reasons for complaint, but to preserve friendly relations with her'.[9] The ministers in Vienna shrank from making any move at all, lest their action should be misconstrued or resented by the French government.

Austria was afraid, said Adair in his memoir, 'that any attempt to re-establish, even for the purposes of peace, a good understanding with other powers, would draw down upon her the vengeance of Napoleon'. She was prepared not to have any foreign policy at all, save that of seeking favour with France. She was ready to take it as the first rule of her conduct that she must avoid too close an intimacy with England.

There were reasons why the Austrians could reconcile themselves to the necessity of keeping England at a distance. The cold reception which Canning had given to their peace proposals after Friedland, the withering way in which he had resolved these into their naked selfishness, and the haughty tone of the comments with which he had surrounded his reply, had created dissatisfaction in Vienna.[10]

> 'Judging from extracts from the despatches to Lord Pembroke', wrote Stadion when the messages arrived, 'we cannot expect much from

> the opinion and good-will of the present English ministry. The instructions... consist of bitter reproaches for the past, unpleasant refusals of demands which we have not made and do not wish to make, and counsels for the future which seem to me rather uncalled for.'[11]

But even if the British government had adopted an attitude more warm and sympathetic, Stadion recognised that there had come into existence a new state of affairs in which it was impossible for England to be useful to Austria in the way that had once been contemplated. Napoleon had made his arrangements in Europe and it was vain to hope that he would alter these for the sake of peace with England. It was not expected that he would now restore the Austrians to their former power, even in return for the recovery of his lost colonies.

> Although it is none the less desirable that there should be peace between the Courts of London and Paris to relieve the present political situation, we can no longer flatter ourselves that the pacification can produce essential changes in what has been agreed upon in the Treaty of Tilsit.

Under these circumstances it was less necessary to cultivate the close friendship of Great Britain. Starhemberg, while ordered to strive to 'keep the court of St James's in good feeling towards us', was instructed to 'abstain for the moment from any proposition which would tend to unite us more or less directly with the present policy of the English Cabinet'.[12] Starhemberg was to adopt the role of observer, to watch carefully over England's relations with her former allies, to note especially the conduct of England in regard to France; but he was to evade any attempt that Canning might make to bring Austria into closer touch with Great Britain. 'You will content yourself with giving general declarations of goodwill.'[13]

It required some weeks for such instructions to reach London from Vienna. Starhemberg in the meantime, like his Russian colleague Alopeus, had come under the spell of Canning. He had found his way to a deeper sympathy with the efforts of England and a closer understanding of her policy than his home government cared to show. While that government complained against the existing cabinet in England, he, in London, would argue like an Englishman and declare that 'every one who desires to see the welfare of England would wish the present ministry to remain in power'. Though he admitted that to some people Canning's attack upon Denmark might seem 'odious', he himself set out its virtues,[14] and repeatedly noted its 'wisdom and foresight', and went beyond many Englishmen in expressing his admiration. To him the Czar's offer of mediation was 'vague and ill-worded', but Canning's reply 'left nothing to be desired'; he could expect no tolerable peace until the secret articles of Tilsit should be divulged. He felt that the mediation of Austria might be more acceptable to the British government than the intervention of the Czar. He saw reason to hope that France might make some overture direct to London or through the Austrian minister in Paris.[15]

But no Englishman could have shown himself at heart more hostile to the idea of maritime peace than Starhemberg. No one could have expressed a clearer conviction of the necessity of continuing the war. His reports bear strong marks of the influence of the English foreign secretary, and it is evident that in the conversations that took place in London Canning's was the ascendant mind. We may take it as the opinion of the British foreign office, when Starhemberg puts forward the view:

> It is Impossible for an impartial observer not to take alarm to some extent at the thought of the harm that may happen to England in the present state of things. It is beyond doubt that the malignity of Napoleon is about to collect its efforts against the country which is the object of his deepest hate; and the most dangerous weapon which he could employ would be the attempt to delude England with the semblance of peace, offered with a disinterested air as coming from a desire for tranquillity.

This, said Starhemberg, would be 'complete degradation', would amount to 'the last act of submission of the whole of Europe'. 'It seems to me that so long as the fires of war burn in some corner of the universe there still remains a hope that one day success may be achieved.' England, he said, had many ways of causing inconvenience and disturbance to France; she had nothing to fear from the continuance of the war, for the idea of an invasion was a 'chimera'; only by making peace and allowing Napoleon to recruit his naval power would she stand in danger of being faced by some future attack.[16] Starhemberg, it is apparent, had surrendered to the whole reasoning of Canning. He, like Alopeus, had given his heart to the coalition. Merfeldt, his fellow countryman at St Petersburg, was in entire agreement with him. He, too, wrote to London to urge the continuance of the war with France.[17]

The Austrian minister in Paris worked more closely to the intentions of his government than either Starhemberg or Merfeldt. After Austerlitz Napoleon, exercising that same felicity of choice as when he proposed Stein to succeed Hardenberg, had fixed upon Metternich for this post. Gentz wrote to this man, for the diversion of historians, 'Your nomination to Paris was a blow to me, and I have not yet been able to recover from it. This soul so pure and elevated as yours ought never to have come into contact with the seat of so many crimes and horrors.' This soul so pure and elevated, picking its way not too disdainfully among the contaminations of the French capital, burned with inner hatred of Napoleon. Metternich handled the policy of Austria with the subtlety and resources of a diplomat born to the technique. It was he who negotiated the treaty of Fontainebleau which settled various questions that had been in controversy between the courts of Paris and Vienna. He did not like the treaty, fought bitterly to secure better terms, and felt that his work needed apology; but whenever he had raised a protest he had been silenced by ugly threats from France. 'I have unfortunately been only too well able to see at every moment of the negotiation that not only has Napoleon ceased to know any bounds, but he has completely thrown off his mask.'[18]

Napoleon for almost a year had been pressing for a treaty of alliance with Austria; it had taxed the ingenuity of the ministers at Vienna to evade the question and postpone such a commitment until hostilities should have come to an end. It was now Austria's turn to be suitor, and it was the work of Metternich to urge her cause. There were rumours that a partition of Turkey had been discussed in the conversations at Tilsit, and it was a point in the policy of Stadion to see that the Ottoman Porte should not be dismembered; but in case Napoleon chose to carry out such a scheme Austria had no intention of permitting the exclusive partnership of Russia in the affair; she was anxious to have a share of the spoils. If the harshness that had been shown in the negotiation for the treaty of Fontainebleau could be taken as an index of the treatment that was to be expected from Napoleon, the court of Vienna could have little hope that France would welcome its proposal of more intimate friendship. It could congratulate itself that, at last, it had settled all doubtful controversies with Napoleon, and that 'no question of any importance was left open'. But, once the idea of an alliance had been tentatively put forward in Paris, it was the opinion of Metternich that 'it is due to our dignity that we should wait until France makes a move'.[19]

Yet behind *arrière-pensée* lurks deeper *arrière-pensée*, and every purpose is lined with mental reservation. In examining the diplomacy of Austria, we can never be sure that we have reached the last filtering, and discovered the final irreducible element. When Tolstoy, the Russian minister, comes to Paris to take charge of the new policy of the Czar, we find that Austria has a still lower stratum of hidden ulterior motive. Metternich fastens himself upon his Russian colleague, assures himself that Tolstoy is at heart a servant of the coalition, and makes the man his confidant. It would seem that there are quiet nudgings and secret whisperings between these two ministers, almost under the very eyes of Napoleon. Metternich will say:

> We all have, and can have, only one end, that of preserving our integrity in the midst of the general overthrow.... The Emperor Napoleon will caress you to-day, only to fall upon you to-morrow; he will do as much with Austria; we shall both of us have to carry on an eternal struggle against his subversive designs.

Tolstoy will reply: 'I am entirely of your opinion.... I do not know what these people wish to do with me, but they are fools if they think that I shall be their dupe.' Metternich's conclusion will be that: 'We have more chances than any other power in Europe of being intact when the great day comes for putting an end to this state of affairs – a state that is essentially precarious, because against nature, and against civilization.'[20] Austria then, is waiting silently, waiting fearfully, for one culminating moment. She is cursed still with an unreconciled heart, and behind all thoughts of affiance with Napoleon she has a remote hankering after a new coalition.

Merfeldt in St Petersburg may walk in the shadow of a British ambassador, and have undignified controversy with the minister of Napoleon, and, in the face of the Tilsit system, show an air of grievance that is out of season. Starhemberg in London may rigidly persist in avowed hatred of Napoleon, and may take his prejudices from Canning until he seems to become a mouthpiece for English official policy. But the diplomats in Vienna will be cold and formal in their relations with England, eager and friendly towards Napoleon, earnestly working for the alliance of France. They are prepared to cover up their true antipathies, and they will only betray their hatred by some evil glance when Napoleon is not looking. Recondite and impassive, proceeding obliquely and always on tiptoe, trusting to subtlety and pure technique, they are men of the *ancien regime*; one can see the handkerchief and the pinch of snuff. In their high abstract world these diplomats are like the giants of the financial market, using a specialised science and terminology, pursuing thought in dry categories of their own, and utterly forgetful of the farmer with his corn, the worker at his loom, the small shareholder at his desk, over whose heads the abstruse speculation is carried on, like a kind of fatality, remote and sinister. Austrian diplomacy works with a point fine as an etching-needle, and is gloriously conscienceless, and is not tortured by a misgiving. It has subtle nuances and courtly euphemisms and polished artifice to cover and cushion cynical purposes. Metternich in Paris serves its ends better than Merfeldt or Starhemberg. He gives it its true tone and flavour. He will secure state secrets by making love to a sister of Bonaparte. And there is an atmosphere in it all that Englishmen come to associate with the name of Metternich, as something malevolent and Jesuitical – the shadow of the Holy Alliance.

This is a game for powdered aristocrats, and it is to be played in full dress. One has pictures of dazzling nights, curtained rooms. One seems to hear, coming from the distance, sounds of music by Mozart. The whole is a world entirely over the heads of the mass of people, working out the web of their destiny like the stars. The policy and the procedure cannot be democratised.

There is nothing communicable to the platform in the diplomacy of the remote *arrière-pensée*. You cannot take the people into your confidence, you cannot evoke national passion as a power behind you, if your immediate object of policy is an alliance with France, and your secret ultimate aim is war. You cannot make a public cause out of a policy of 'mental reservation'. So it is useless to expect from Vienna a significant and revealing gesture such as Canning would give at times; the last secret of Austrian policy is rather to be caught in an almost imperceptible flicker of the eye that conveys something from Metternich to the Russian Tolstoy, when Napoleon's back is turned. The rulers of Austria were unwilling to make the people their accomplices. They knew too well that if they made war itself national and popular, and called upon their subjects to help them against the French, they would be raising at their very doors a monster that would be more terrible to them than Napoleon himself. They were not yet reconciled to this.

Such was the situation, and attitude and character of the power which Napoleon, after he had heard of the expedition to Copenhagen, decided to call out of its persistent neutrality into the war against England.

In October 1807 Napoleon was in violent mood. Portugal had dared to show defiance; England had made the surprise attack on Denmark. Man of giant angers, the Emperor did not know how to meet a thwarting, and made an undignified outburst in public. Lord Whitworth had seen such a display in 1808, had found Napoleon 'evidently under very considerable agitation', speaking 'loud enough to be overheard by the two hundred people who were present', and conducting himself 'with the most extreme impropriety... the total lack of dignity as well as of decency'.[21] Kleist had had a similar experience during the recent campaign; Napoleon 'really had the appearance of being greatly disturbed; and it made him distracted and caused him to keep on repeating the same thing'.[22] Metternich was to be confronted with the same violence at Dresden in 1818 – dramatic boasts, explosive threats, wild ejaculations of a man who 'allowed his anger to carry him away'.[23] Napoleon found it useful to make his name terrible to Europe, and had a policy of intimidating diplomats; but it is a mistake to regard such demonstrations as always consciously contrived; they belong to an unreclaimed tract in the personality of the Corsican, the something ungovernable in his nature, a wildness at the seat of his mind. These noisy and petulant outbursts would recur at times to shock a well-bred Talleyrand, and ruin the impressiveness and the fine finish that had been upon the courts of more royal days. They betrayed a weakness or distress of Napoleon himself in the contingencies of the hour.

On the 16 October 1807 Metternich in a despatch from Paris describes such a dramatic moment. Napoleon at an audience exploded into 'one of the most violent outbursts that he has ever made in a diplomatic circle'. The minister of Portugal received the first shock. 'If Portugal does not do what I want', said the Emperor, 'the house of Braganza will have ceased to reign in Europe within two months from now.'

> I will not allow a single British envoy in Europe; I will declare war against any power which has one two months from now. I have three hundred thousand Russians at my disposal and with a powerful ally like this I can do anything. Since the English have declared that they will not respect neutrals on the water any longer, I will not recognise them on land.

He passed on to the ambassador of Etruria. 'Your Queen is in secret connection with England – but I have settled with that.' To a deputy from Bremen he asked, 'How are things going with you?' The deputy said, 'Badly sir.' 'Very well,' was the retort, 'they will go worse even yet. Bremen and Hamburg are English towns; I shall know how to treat them for this.' Metternich did not attempt to describe the impression these exclamations made on all the people present, or the violent manner in which they were uttered.[24] For this episode was merely the prelude to an incident that more nearly concerned Austria.

An hour after this disconcerting experience Metternich had an interview with the French foreign minister. Champagny had unpleasant revelations to make. The Emperor, said he, would be satisfied with half-measures no longer. England had refused the mediation of Russia – she must be compelled to make peace.

> Only three powers are left in Europe; France, Russia and Austria. It is for you to contribute directly to the salutary work which the Emperor is undertaking. You have had reason to see for a long time that your flag is not respected any more than the others; the King [of England] has just declared that he will recognise no neutrality.... Russia is going to make common cause with us. All Italy obeys the Impulse from France. There remains only you to close all access to the Continent.... You, more than any other power, are interested in seeing peace and tranquillity restored, but this is impossible without the maritime settlement.... The Emperor wishes you to say that if England does not give up the Danish fleet and revoke the principles enunciated in her latest declaration, you will recall your minister from London and at the same time – say the first of December – dismiss the English minister from Vienna.

Metternich protested against this demand. The Austrian ports were already closed to English vessels; no good would be achieved by making a definite declaration of war; and much harm would result to Austria. Metternich asked how it could happen that Napoleon could make this demand and yet when the Austrians had proposed that their 'relations of friendship and good understanding should be grounded in the same way upon more decided bases'; he had given no answer, as though he did not desire a more intimate alliance.

Metternich, though he protested against the demand, could not recommend his government to resist, but wrote:

> Every reflection on the communication which the present report puts forward, seems to me to be rendered superfluous by the positive fact that Napoleon, scorning all moderation and even policy, *has made it understood that* he would accompany the *official demand... with a declaration of war,*

> *in the event of our refusal. I can vouch for this assertion.... There has recently been a total change in the methods of Napoleon; he seems to think that he has reached a point where moderation is a useless obstacle.* The peace of Tilsit and the extreme weakness of the Emperor Alexander, could not but bring him to this point.... Of all the continental powers we are the one most directly and immediately menaced.... On this occasion more than on any other the fury of Napoleon will have no limits; and his thirst for universal dominion could not be better served than by our coming short of satisfying his desires.[25]

The court of Vienna which had already resolved to make no hesitation in answering to the wishes of Napoleon, and indeed had no intention of braving the fury of France, did not remain indifferent to the force of Metternich's words.

In the last week of November Starhemberg in London received an angry despatch from Vienna. It was dated the 27 October and had been sent to England via Paris so that it might come under the eyes of the French government. It began by enumerating the successive attempts which Austria had made to promote the peace of Europe; it deplored the failure of the mediation that had been offered before the battle of Friedland and complained of the unfavourable welcome which Canning had given to the further overtures that had followed the battle. By rejecting the good offices of Russia, and by making the attack on Denmark – an outrage which indeed had never been satisfactorily accounted for – England, said Stadion,

> seems to have announced far and wide, her intention of removing all possibility of a pacification, and of refusing to do anything to restore tranquillity to Europe... and if the British court has not once and for all renounced any thought of uniting and Identifying her interests with the cause of Europe, It ought to be convinced on its own account of the disastrous consequences of this conduct.

It was impossible to calculate the evils which would result from the policy of Great Britain, the despatch went on to say; but things had reached the point at which the court of Vienna felt itself compelled 'to insist more peremptorily than ever on learning the real intentions of the Cabinet of St James's in this matter, and on impelling it if possible to sentiments more just and more conformable to the situation'. Starhemberg therefore was to address himself 'for the last time, and officially', with a 'formal proposition' to the English ministry. He was to secure a sincere declaration of England's willingness to negotiate for peace; and 'as a preliminary guarantee of its intentions the British government was to renounce completely the hostile measures recently taken against Denmark and to cancel the declarations that accompanied these'.

> 'You will feel the Importance of the commission which has been entrusted to you', wrote Stadion. 'There is no reasoning which you ought not to employ, no argument which should be left unused to make the English Ministry appreciate the only point of view in which this proposition can be envisaged. If on one side the welfare of Europe depends to a great degree on the decision which the British Cabinet will choose to take in these circumstances, its own position equally demands that it should at last put an end to a struggle... which will inevitably involve the destruction of England herself.'

If the British government should refuse to give satisfaction on these questions or should provoke deliberate delays, then Starhemberg was to ask for his passports and Austria would break off relations with England.[26]

An accompanying despatch of the same date recounted the private grievances of the court of Vienna against the British government. It referred to the injuries done to Austrian shipping in defiance of neutral rights. It complained of the ineffectiveness of Austrian protests that had been made periodically against the injuries, 'The Court of London has for nearly a year treated us in this matter as though we were veritable enemies.' If the British government showed itself reasonable on the question of peace, Starhemberg was still to press this maritime issue, and was to demand satisfaction.[27]

All this was in the language of unmistakable hostility, and was calculated to provoke rather than to persuade. Nothing in it bore the mark of a sincerely mediating power, and everywhere there seemed a disposition to raise points of dispute rather than to smooth them over. But a secret and cyphered despatch was drawn up for Starhemberg at the same time, and was not sent via Paris at all, but was transmitted by a more direct route. This revealed the true situation of affairs.

> We have had to choose between instant war and the measure we have just taken. The cabinet of London might have spared us a step which costs as much to the heart as to the judgment of His Imperial Majesty, if it had chosen to enter with more friendship and more consideration for the well-being of Europe, into the repeated representations which we have been making for six months. As things are at present we have had to give way to the necessity of the moment. If I have gathered the real meaning of the last advice which M. Canning charged Lord Pembroke to give us I am assured that that Minister foresaw in some manner the extremity in which we were soon to find ourselves, that he felt that we ought in such an event to make this last submission and that he saw it was Important to Europe and to England that our existence should be preserved intact in the midst of this universal political overthrow.[28]

Stadion at the same time explained to the English minister Adair 'that no concession which His Imperial Majesty had been called upon to make during the whole course of his reign, affected him more deeply than the measures which had been extorted from him upon this occasion'.[29]

Starhemberg was alarmed when he received the commands from Vienna. He had approved of the reply which England had given to the Russian mediation; he had applauded the 'wisdom and forethought' of the Danish expedition; he had declared that it would be disastrous for Europe if Canning should make peace with France. He sought counsel of his Russian colleague Alopeus,[30] and the two put heads together. They decided that the Austrian overture, presented in the hostile terms in which Count Stadion had drawn it up, would only defeat its own ends.

> 'I know too well', wrote Starhemberg, 'the sort of pride and even of violence which always characterises the English Ministers of whatever party, and I could not fail to see how necessary it was not to give them a direct affront.... The faintest show of menace would have destroyed all hope at the very start.'

Starhemberg's official overture to Canning therefore was embodied in a note that omitted any point likely to cause offence.[31] In particular, it made no mention of Denmark.

> I did not hide from [Mr Canning] at the same time in my conversations, that we were making the demand of reparations to Denmark an essential point.... The verbal reply of the Secretary of State 'that If negotiations were commenced they would bring up the question of Denmark, which would enter into the general settlement', seemed to me too definite and too sincere for me to lay stress In my Note upon an object which at the present moment would have given the signal for an Immediate rupture. It would have been considered that we wished to dictate terms in an imperious manner.[32]

Canning in reply to the vague note of Starhemberg merely renewed his previous protestations of a desire for peace. He affected to regard the overture as a repetition of Austria's previous proposals.[33] But the court of Vienna had distinctly declared that the king of pious assertions, which Canning had hitherto given, would by no means satisfy its demands; and since its minister in London had already taken the liberty of diluting the sense of his instructions and softening the peremptory nature of the commission that was entrusted to him, it was not to be expected that that court should feel contented if Canning on his side failed to respond to the friendly action and produced the very reply that had been declared insufficient. The proper thing, the expected thing, at this point, therefore, was that Starhemberg should at least follow out the remainder of his instructions and meet Canning's evasiveness by breaking off relations with England.

Instead of this Starhemberg acted again on his own responsibility. He declared that Canning's reply

> while not coming near to what I would desire, seems to me more satisfactory than I had dared to hope.... It seems to me that it is possible to perceive in these expressions, if one looks for them, a real opening and a disposition to listen to peace proposals.

The British government was insisting that before negotiations should be opened, the French should make an announcement of the principles upon which they would consent to treat. This was a point which the Czar himself had consistently demanded until the battle of Friedland had made him rush into the arms of Napoleon. It represented no new concession on the part of England, no departure from previous policy. Starhemberg feigned to see in it a ray of hope, and nothing more than this was needed to make him once more disobey the commands of his government. He deferred the breaking off of diplomatic relations, and took no steps to withdraw his mission from London. He decided first to communicate with Metternich in Paris and discover whether Napoleon would meet the English demands. He asked the French to make an avowal of the principles upon which they would consent to negotiate. To his government he sent his excuses:

> It is below the dignity of a great court to bring the affair to a standstill on a mere matter of *punctilio* – a question of who shall make the first overtures. The interests involved are too great for us not to make every attempt to attain our object... I take the liberty of observing to Your Excellency that If France should now make just and reasonable offers to Great Britain, and this power reject them with contempt, all Europe would see the real designs of England, and the present ministry would be absolutely unmasked.[34]

Starhemberg sent a duplicate of this despatch to Metternich in Paris, and Metternich, after making careful modifications,[35] posted a copy, together with Canning's note, to the French foreign minister who was with Napoleon at Milan.

> 'His Majesty the Emperor Napoleon', wrote Metternich, 'will in his great wisdom decide if he thinks he can discover in the British Cabinet's reply, any motives for following up this opening. Nothing but the hope of being useful to the cause of peace has made our envoy put off the declaration of a rupture in the official form, such as he was instructed to give. He awaits precise directions from me. I in my turn ask for those of His Majesty the Emperor of the French, and am convinced that by this new mark of confidence I am entirely fulfilling the views of my court.'[36]

It becomes evident at this point that, like Starhemberg in London, Metternich himself is anxious to see the conclusion of peace between England and France, in order that his country may be liberated from the dilemma in which it is placed – in order that it may avoid the declaration of war which Napoleon has ordered to be made against England.[37] But the government of Austria, like that of Russia, is ready to move more quickly than its servants abroad, is more anxious to prevent anything that might give the faintest shadow of offence to France, and is more disposed to hasten the quarrel with England. This is illustrated by the reproof which Starhemberg receives from Vienna, when it is learned that he has taken it upon himself to disregard the extreme tenour of his instructions. The Emperor of Austria is so severe that Tolstoy, the Russian minister in Paris, is soon found declaring: 'it is a thing certainly not out of keeping with all the queer antics of fate in this century to see France checking Austria in her resentment against England'.

The Austrian government was greatly perturbed by the conduct which its minister Starhemberg had taken upon himself to follow. It did not imagine that Napoleon was anxious to make peace or to have the question of a negotiation so pressingly pursued at London. It had had from Metternich the report of the warlike and intemperate language which the French Emperor was using in regard to England. It had been anxious to show Napoleon its readiness to join his continental policy and break off relations with the court of London – had ordered Starhemberg, in case England gave satisfaction on the prime question of a negotiation, to pick a further quarrel on the private maritime grievances of Austria. Much as Austrian interests might require the conclusion of a final peace, Starhemberg, by taking his pacific role too seriously, had compromised his government, had made it appear that his country was a slow and unwilling partner in the confederacy against England. He had put the court of Vienna in the danger of receiving a swift rebuff from France.

> 'Once it was determined to transmit an official Note on this subject, this diplomatic document ought of necessity to have comprised the whole of the demands which you were instructed to give', wrote the Austrian chancellor.
>
> 'Mr Canning had a right to say in his reply that what Your Excellency is insisting upon in your Note is nothing but a repetition of our former overtures on the same subject, and that he therefore could only repeat ... the same declaration he made to you at that time. ... And you, Prince, must recognise for yourself that a note from that Cabinet, merely a duplicate of those which England has already addressed to us and to Russia, could not be in any way satisfactory.'

Unless Napoleon should be willing to follow up the overture, unless England should make a more explicit declaration, and unless the French court should signify through Metternich its desire to see Starhemberg prolong his stay in London, the minister was to open the question straight away, and in an official manner, to Canning, and follow literally the instructions which had been sent to him at the end of October. 'His Majesty also desires you to mention about the demand of passports in the official Note... or simultaneously with it, at least; and he wishes you to allow no delay in replying, beyond two or three days.'[38]

An accompanying despatch ordered Starhemberg to follow the counsels of Metternich in future; Vienna was too far distant to allow for communication. If the improbable thing happened and France really wished to renew the discussions with England, then Prince Starhemberg 'must use all possible zeal' to promote the negotiation, following all the time the directions of Metternich.

> We have too keen an interest in seeing the re-establishment of general tranquillity not to desire such an event. The Emperor commands... that you make a point of fastening with zeal and activity upon everything which might provide a nucleus for a real rapprochement between the two belligerent Powers.[39]

This reproof was sent to Starhemberg via Paris, under a flying seal. Metternich showed it to the French government. The Russian minister in Paris saw the meaning of the severe language of Austria; Stadion had taken this course 'probably with the intention of making capital out of it here'.[40]

> 'Napoleon', said the French minister, 'applauded the wise and strong conduct that the Court of Vienna had followed on this occasion; it did the greatest honour to the Emperor of Austria; the Prince of Starhemberg, having dared to forgo the execution of the orders of his Court, deserved the reprimand.... But it might be detrimental to the success of the affair for Starhemberg to be turned from the course he had followed; and the Emperor believed it would be more useful to avoid upsetting his point of view in the matter, by preventing despatches which might throw him into doubt, from reaching him.'[41]

The French minister, writing to Metternich from Milan, had already disclosed the intentions of Napoleon. Comparing the reply which England had made to the latest Austrian overture with the answer recently given to the Czar's offer of mediation, he found the 'same method of evasion', he saw 'the desire to gain time and to postpone an Austrian measure that is calculated to cause a sensation in England by no means favourable to the present ministry'. Canning had said that His Britannic Majesty was as willing to negotiate with France as he had always been, that a fresh declaration on this matter was unnecessary, but that he was quite willing to give it.

> 'But did England desire peace when she rejected the mediation of Russia?' wrote Champagny.'... The English ministry does not say it wishes for

peace. It makes its willingness to enter into negotiations depend on a thousand conditions; it desires a peace which will establish on equal bases the respective interests of the powers taking part in the war, a peace conformable to the fidelity which England owes to her allies (has England an ally in Europe now?) – a peace which will provide for the security and tranquillity of Europe.... When England expresses her desire for peace we are therefore entitled to ask for proofs. In the present situation, let her prove her sincerity by naming plenipotentiaries to treat for peace with France and Russia. Prince Starhemberg might in that case remain in London, and he will be authorised to deliver passports to these ministers, so that they may cross over to the continent.'

The whole French declaration was drawn up in a manner hostile, provoking and proud. Napoleon did not break off the discussion; he allowed everything to depend upon the next movement of the British government; he was determined to put England in the wrong; that is why he threw upon Canning the burden of declining to send plenipotentiaries to Paris. The note stated that if Austria would join hands with those powers which had declared against England, this step, 'taken in the same month as that of the emperor of Russia, would be the very thing to scatter perfidious hopes and open the eyes of the English people to its true situation'.[42] The edict of Milan, dated the 17 of December and containing a renewed attack on English trade and on the remnant of neutrality that still existed in Europe, confirms the idea that at heart Napoleon had no desire for peace, no intention of making concessions that would conciliate the court of London. Metternich spoke truly, in transmitting the French note to England, when he said that Canning would have to be very careful in his choice of plenipotentiaries.

Metternich was anxious to make it plain to Napoleon that Austria was completely entering into the views of France. When he dispatched the French note to London, he sent to Starhemberg a statement of his own views, taking care to catch the mood of the French government, and writing in an unfriendly strain.

'It seems to me essential', he declared, 'that when you acquit yourself of the honourable commission with which you are charged you should at the same time declare positively that the refusal to send negotiators furnished with full powers sufficient for treating, will be the signal for the immediate demand for your passports.'[43]

But he accompanied this ostensible and official instruction with a private and secret letter that was not allowed to come under the eyes of the French. 'Your Highness will see at first reading', it began, 'that my despatch of the 28rd December is drawn up in an offensive manner. It was in fact intended to be shown to M. de Champagny.'[44] In this more intimate letter we are justified in finding the real sentiments of Metternich on the subject of maritime peace; for his arguments are earnest and cogent, and they dovetail into the whole system of his views at this period. Metternich was anxious to see peace in Europe for the moment, in order that Napoleon might have no pretext for further exploitations of his military supremacy. His desire was to see the powers acquiesce in the settlement of Tilsit, and end the universal chaos by making an interval in the contest; but it was only a case of stooping to conquer, the enemies of Napoleon would be quietly gathering power, and some future contingency would call out a coalition once more. It was in conformity with this design, this last lurking *arrière-pensée* of the Austrians, that Metternich, writing privately to Starhemberg at this moment, should press England to make peace – a peace riddled, it is true, with mental reservations, and resting on secret hostility – and should use every argument to prove that this policy was for the enemies of France the highest opportunism. The fact that after the expedition to Copenhagen, Napoleon showed himself little disposed for a settlement, seemed only to confirm Metternich in his fundamental idea that the event most disadvantageous to France in the contingencies of the moment would be the conclusion of a maritime peace which should rescue England from her dangerous isolation, and should disarm Napoleon at a time when his power seemed irresistible and so should relieve the burdens of the European states. Thus we find Metternich writing:

> Let the British ministry think seriously about the step it is taking; the bows are drawn as tightly as possible; few of the states of the old Europe have escaped the prolonged scourge of the maritime war; two of them, only, remain strong, powerful, intact. Let England not wait for new complications and evils which are too easily to be foreseen and which will weaken the last supports of the old order of things. Let her make peace or at least negotiate. Months gained now are worth years at any other time or under other circumstances.

Many things, thought Metternich, made the moment opportune. The Russians 'by every motive of commerce and existence' were interested in seeing the establishment of peace. England would find in Austria an intermediary whose sole desire was for the immediate establishment of a tolerable if not a stable condition of things.

> When will a negotiation with France be renewed under such auspices if the present opportunity is allowed to slip?...
>
> Let the English ministry above all things not bring the affair to a halt on preliminary questions, such as the announcement of bases prior to the assembly of negotiators. It will be Impossible to defend your position in

> London, Prince, if in place of the sending or the positive announcement of the arrival of negotiators, there is a mere desire to continue pourparlers.[45]

At a later date Metternich added in a further private and secret despatch: 'The British ministry will be very foolish if it does not lend itself to a negotiation.'[46] Starhemberg, in the meantime, had been working in London to prepare the way for further measures. He was pleased to think that at least his conduct had not met with impetuous reproof from Napoleon. 'The couriers delay gives hope that he will bring overtures from France, and I am bold enough to flatter myself that these will not be rejected if they are at all reasonable.' Canning, indeed, had said 'that a counter-proposal would certainly be sent from England if the French proposals gave any sign of there being a real desire for peace'.[47] It now became the business of Starhemberg to be the apologist of the French proposals, to point out how acceptable they were, and to make a pathetic fight for a cause against which he had recently written with so much eloquence and conviction.

Between this man and the English foreign secretary there ensued a wearisome argument on the question of a negotiation with France. It quickly developed into an unpleasant kind of duel with futile fencing and evasiveness on both sides. Before it was finally abandoned it had produced irritation between the two ministers, and even something that seems like an estrangement.

Metternich's recondite policy of peace with mental reservations, and Austria's cool cautiousness, were not the course that would appeal to Canning, hasty emotional man, 'all nervous and a-quiver'. The treaties of Tilsit had found him masterly and unrelenting; in the mediation of Russia he had detected veiled hostility and even direct deception; he was not thereafter disposed to put faith in vague peace moves engineered from Paris. He complained, as he had always done, that France persisted in refusing to avow the basis upon which she would negotiate. It did not placate his fiery heart when Starhemberg argued that this was a proof of Napoleon's good faith, that France was merely unwilling to raise preliminary discussions which might wreck the negotiation before the plenipotentiaries had come together. The Austrian minister described Napoleon's eagerness for peace, declared himself ready to provide passports for English plenipotentiaries and communicated with members of the opposition, in order to reinforce his case.

> 'I cannot omit to say', he reported, 'that the secretary of state has appeared to me to be struck at times with the force of my arguments, and even embarrassed for a reply to my redoubled objections... and I have noticed him resorting to the most weak of pretexts to account for his refusal.'

But impatience and unpleasantness resulted from this argumentation. It became dangerous for Starhemberg to remain in London, if he could have nothing with which to justify the course he was pursuing. On the 1 January he made official communication of the proposals he had been authorised to make; drawing up the note, as he informed his government, in the most moderate terms that were possible, 'in order not to ruin everything by too much energy or hastiness, and 'to put the English ministry completely in the wrong before Parliament'.[48]

On the 8 January Canning wrote a reply which the irritated Austrian called 'very long, very diffuse, and very ungracious, as well as badly drawn up'.[49] He declined to avail himself of the opening that France had given. Napoleon had refused to declare the principles upon which he intended to negotiate. A hostile capital, like Paris, was an unsuitable place for the meeting of plenipotentiaries. England would not make any arrangement with France unless her allies should be comprised in it. Finally, the overture was defective in form.

> 'The overture of Prince Starhemberg…', wrote Canning to Adair, '…made by him without the knowledge or sanction of his court, and simply on a suggestion of one of the French ministers, which is believed to have been conveyed in a rather unofficial form to Count Metternich at Paris.'[50]

The Austrian minister, having received the English reply, immediately wrote to ask for an interview with the foreign secretary. Canning, at this time in a bad state of health, was away in the country when the letter arrived; but the undersecretary Hammond had been instructed to open his correspondence, and he replied to the Austrian. 'Mr Canning', he wrote, '…considers the subject which he has been discussing with you to be of so delicate and important a nature as to render it desirable that all communications respecting it should be made solely in writing.'[51] Starhemberg gave the appearance of becoming more and more irritated. He decided to make it known verbally and unofficially to the British government that he proposed very soon to ask for his passports. Before doing this he played his last card, and wrote to Metternich:

> If at the present moment you could send me propositions and a basis of peace to offer to England, together with an authorisation for my conduct such as would silence all objections, the ministry, already embarrassed for excuses for its present conduct, could not go on refusing to enter Into negotiations. It would be worthy of the generosity of Napoleon for him to make this last offer for the general good of Europe.[52]

But Metternich was now as peremptory as his government, requiring the instant departure of the Austrian mission from London, if England did not accept the clear alternative of sending plenipotentiaries to Paris. On the 16 January he made it known that Napoleon expected the final break with England to take place by the 21 January. Before this announcement arrived, Starhemberg had abandoned his efforts and had decided to leave England.[53]

He was not convinced, even yet, that the possibilities of the situation had been exhausted. He was still obsessed with the idea that peace was the superlative object, and that his business was to promote a negotiation. But perhaps he was conscious of having lost favour with Napoleon, and desired to restore his credit; for his arguments took more and more a turn hostile to the British ministers. He asserted that 'if France had seriously desired it' she could have induced the English ministry to negotiate, and 'might still do so to-day'.' He cited Canning's own words as his justification:[54]

> As soon as the Basis of negotiation shall have been satisfactorily ascertained and an unexceptionable Place of Negotiation agreed upon, His Majesty will be prepared to name Plenipotentiaries to meet those of the other Powers engaged In the War; but his Majesty will not again consent to send his plenipotentiaries to a hostile Capital.

But even if Canning should prove still intractable, Starhemberg continually played with the idea of 'forcing the ministry to treat'. He had been careful to communicate his overtures to members of the opposition in England – to Grenville, Holland, Grey and Sheridan. He had noted the embarrassment of Canning at times, and had talked of 'having the ministers completely unmasked'. He declared: 'The opening of parliament provides the means of destroying the only obstacle, and annihilating the prejudice that the present ministry has inculcated throughout the country, that Napoleon does not seriously wish to see the end of the war.'[55] He announced his intention of pointing out all this to the French Emperor, if he could secure an interview as he passed through Paris: Metternich in the same way had talked of 'calling the English ministry before the tribunal of public opinion'. The Austrians and the French at this period were paying watchful attention to the stability and strength of the existing cabinet in London. In the French foreign office archives there is mention of a scheme to offer England a treaty of commerce in order to seduce the British traders and place them in opposition to the warlike policy of their government.[56] Napoleon had good reason to work for the overthrow of Canning.

Starhemberg sent a long justification of his conduct to his government at Vienna, explaining how little he had departed from the intentions of his government.[57] It is interesting to see that he called Alopeus – of all people – to bear witness to his good policy. Certainly his decision to depart from the original tenour of his instructions was made after consultation with this man.

'I thought it best to give him every confidence', wrote Starhemberg; and the old intriguer did not cease to assist with his 'advice and good offices', even after the Russian government had ordered him to break off diplomatic relations with England. He declared that he would have disobeyed the positive instructions of the Czar and would have withheld the declaration of war, in order to be able to lend official support to the Austrian overture, but that he was prevented from taking this course, since the British government had already learned of the rupture. It was Alopeus who counselled moderation and patience; advised Starhemberg, even at the end of the negotiation, to postpone the demand for his passports; and even recommended him to make some delay before using them, when they had actually been obtained. 'Perhaps he would have preferred me not to leave London so promptly as I did.' When the harassed Austrian at a perplexing moment writes to his government that he 'does not know how to be thankful enough' for Alopeus, we who know the wily Russian can conjecture what had taken place; and we can see it as all the more significant in the light of one final episode which remains to he examined – and in which Alopeus is the central figure.

Notes

1. Andréossy to Talleyrand, 18 July 1807, in *A[rchives du ministère des] A[ffaires] É[trangères] [Paris], Autriche*, t. 880.
2. G. Jackson to G. Canning, No. 4, 20 July 1807, *F. O. Prussia* 64/75. See also Alex. Wassiltchikow, *Les Razoumovski*, II, 3me partie, p. 131.
3. Andréossy to Talleyrand, 24 Aug. 1807, *A.A.E. Autriche*, t. 380.
4. Soltikov (acting foreign minister in St Petersburg) to Budberg, 25 June/7 July 1807, in *Sbornik*, LXXXVIII, 74 – 6.
5. Savary to Napoleon, 6 Aug. 1807, in *Sbornik*, LXXXIII, 7 – 10 and *passim*.
6. Stadion to Starhemberg, 6 Sept. 1807 (No. 4), *W.S.A. Weisungen nach London*.
7. Starhemberg to Stadion, No. 63, 8 Aug. 1807, *W.S.A. Berichte aus London*. G. Canning to Lord Pembroke, No. 16, 17 Aug. 1807, *F. O. Austria* 7/85.
8. Duroc to Andréossy, 18 July 1807, *A.A.E. Autriche*, t. 380. Talleyrand to Andréossy, 3 Aug. 1807, ibid.
9. Stadion to Starhemberg, 6 Sept. 1807 (No. 1), *W.S.A. Weisungen nach London*.
10. Adair. op.cit., p. 264. Cf. Stadion to Metternich, 3 Aug. 1807, quoted in Beer, *Zehn Jahre Oesterreichischer Politik*, p. 296.
11. Stadion to Starhemberg, 6 Sept. 1807 (No. 2), *W.S.A. Weisungen nach London*.
12. Stadion to Starhemberg, 6 Sept. 1807 (No. 3), *W.S.A. Weisungen nach London*.
13. Ibid.
14. Starhemberg to Stadion, No. 65, 2 Sept. 1807, *W.S.A. Berichte aus London*.
15. Starhemberg to Stadion, No. 66, 4 Sept. 1807, ibid.
16. Starhemberg to Stadion, No. 65, 2 Sept. 1807, *W.S.A. Berichte aus London*.
17. Caulaincourt to Napoleon, No. 115, 12 June 1808, in Grand-Duke Nicholas Mikhaïlovitch, *Relations diplomatiques de la Russie et de la France*, 11, 186.
18. Metternich to Stadion, 12 Oct. 1807, in *Mémoires du Prince de Metternich* (2nd edn 1880), II, 125.
19. Metternich to Stadion, 12 Nov. 1807, in *Mémoires du Prince de Metternich*, II, 128.
20. Metternich to Stadion, 12 Nov. 1807, in *Mémoires du Prince de Metternich*, II, 136 – 9. Cf. *Sbornik*, LXXXIX, pp. 252 – 5. Tolstoy to Romantsov (Particulière et secrète), 27 Nov./ 9 Dec. 1807.

21. Lord Whitworth's dispatch of 14 Mar. 1803, in O. Browning, *England and Napoleon in 1803*, pp. 115 – 117.
22. Bericht von Kleist, 2 Mar. 1807, in Bailleu, *Preussen und Frankreich*, 1795 – 1807, II, 586 – 9.
23. *Mémoires du Prince de Metternich*, I, 146 – 54; II, 4a61 – 3.
24. Metternich to Stadion, 16 Oct. 1807, in *Mémoires du Prince de Metternich*, t. II, pp. 129 – 30.
25. Metternich to Stadio, 16, Oct. 1807, in *Mémoires du Prince de Metternich*, t. II, pp. 128-36.
26. Stadion to Starhemberg, 27 Oct. 1807 (No. 1), *W.S.A. Weisungen nach London.*
27. Stadion to Starhemberg, 27 Oct. 1807 (No. 2), *W.S.A. Weisungen nach London.*
28. Stadion to Starhemberg, 27 Oct. 1807 (No. 3), *W.S.A. Weisungen nach London.*
29. R. Adair to G. Canning, Separate, 25 Nov. 1807, *F. O. Austria* 7/84. Printed in Adair, op.cit., p. 291.
30. 'As I know for certain', wrote the Prussian minister in London. Jacobi an den König, 29 Dec. 1807, in Hassell, *Geschichte der Preussischen Politik* 1807 *bis* 1815, p. 356.
31. Starhemberg to Canning, 20 Nov. 1807, in *Parliamentary Debates*, X, col. 104, No. III.
32. Starhemberg to Stadion, 27 Nov. 1807, *W.S.A. Berichte aus London.*
33. Canning to Starhemberg, 23 Nov. 1807, in *Parliamentary Debates*, x, cols 104 – 5, No. IV. Cf. G. Canning to R. Adair, 19 Jan. 1808, *F. O. Austria* 7/84.
34. Starhemberg to Stadion, 27 Nov. 1807, *W.S.A. Berichte aus London.*
35. The dispatch is printed (with marginal notes of the alterations which Metternich made) in *Sbornik*, LXXXIX, 287 – 9.
36. Metternich to Champagny, 4 Dec. 1807, in *Sbornik*, LXXXIX, 286 – 7.
37. Tolstoy to Romantsov, 29 Dec. 1807/10 Jan. 1808, in *Sbornik*, LXXXIX, 324.
38. Stadion to Starhemberg, 22 Dec. 1807 (No. 1), *W.S.A. Weisungen nach London.*
39. Stadion to Starhemberg, 22 Dec. 1807 (No. 2), *W.S.A. Weisungen nach London.*
40. *Sbornik*, LXXXIX, 324.
41. Tolstoy to Romantsov, 30 Dec. 1807/11 Jan. 1808, in *Sbornik*, LXXXIX, 331.
42. Champagny to Metternich, 15 Dec. 1807, in *Sbornik*, LXXXIX, 300 – 3.
43. Metternich to Starhemberg, 23 Dec. 1807, in *Sbornik*, LXXXIX, 306 – 7.
44. Metternich to Starhemberg, 26 Dec. 1807, in *Sbornik*, LXXXIX, 308 – 10.
45. Metternich to Starhemberg, 26 Dec. 1807, in *Sbornik*, LXXXIX, 308 – 10.
46. Metternich to Starhemberg, 10 Jan. 1808, in *Sbornik*, LXXXIX, 335.
47. Starhemberg to Stadion, 22 Dec. 1807, W.S.A. *Berichte aus London.*
48. Starhemberg to Stadion, 5 Jan. 1808, W.S.A. *Berichte aus London.* Starhemberg's note to Canning (1 Jan. 1808) is printed in *Parliamentary Debates*, X, cols 105 – 6, No. V.
49. Starhemberg to Stadion, 10 Jan. 1808, *W.S.A. Berichte aus London.* Canning's note is printed in *Parliamentary Debates*, X, cols 106 – 8, No. VI.
50. G. Canning to R. Adair, 19 Jan. 1808, *F. O. Austria* 7/84.
51. Hammond to Starhemberg, 9 Jan. 1808, *F. O. Austria* 7/87.
52. Starhemberg to Metternich, 10 Jan. 1808, *W.S.A. Berichte aus London.*
53. Starhemberg to Stadion, 21 Jan. 1808, *W.S.A. Berichte aus London.*
54. Starhemberg to Stadion, 4 Feb. 1808, *W.S.A. Berichte aus London.*
55. Starhemberg to Stadion, 26 Jan. 1808, *W.S.A. Berichte aus London.*
56. Réflexions sur la Paix avec l'Angleterre, *A.A.E.Angleterre*, t. 604.
57. Starhemberg to Stadion, 4 Feb. 1808, *W.S.A. Berichte aus London.*

5
Crowe's Memorandum of 1 January 1907*

I said at our January meeting that, if I could find the time I should like to circulate a commentary on part of Desmond Williams's paper on 'The Role of National Prejudice in the Formation of Foreign Policy'.[1] The note need not engage any of the Committee's discussion-time, but is intended to carry rather the implication that part of the Committee's discussion may take place on paper.

Some of the reflections which Desmond Williams's essay provoked in me really have further reference to William Armstrong's paper on 'The Nature of the Decision-Making Process',[2] and, still more particularly, to things said in the discussion subsequent to that. These latter points I have tried to deal with in the second part of my paper.

I History and policy

An important section of Desmond Williams's paper on 'The Role of National Prejudice in the Formation of Foreign Policy' raises the question of Crowe's famous memorandum of 1 January 1907[3] as well as Lord Sanderson's counter-memorandum of 21 February 1907.[4] The comparison of these two documents raises issues highly relevant to our Committee, but it is not clear that these are usefully formulated when stated in terms of 'national prejudice', as this is usually understood. If we grant that Crowe's memorandum suffers from errors or aberrations or prejudice, we are likely to find that the mathematical formula for these (or the cause of the deflections that occurred) involved nothing that was peculiarly British at all. The papers of Crowe and Sanderson represent two ways of envisaging policy which may divide statesmen within any country – they are (by accident) British examples of a dichotomy which is universal – and each of the modes of operation is liable to be as much the cause as the effect of the prejudices that come to be associated with it. If Sanderson seemed kinder to Germany than Crowe, this can hardly have been the result of a notorious predilection for that country. Sir Charles Hardinge, commenting on his counter-memorandum on 21 February 1907, writes: 'Somewhat to my surprise he has taken up the cudgels for Germany.'[5]

A primary difference between Crowe and Sanderson can be described as 'historiographical'; and twenty-seven out of the twenty-eight paragraphs in the counter-memorandum relate to the interpretation of past events. Here the difference corresponds to a remarkable degree with that which exists between diplomatic historians, some of whom are excessively 'national' while others reach a more genuinely 'international' point of view. Those who write diplomatic history from documents in the British Foreign Office, and from these alone, tend to be too self-regarding and show a certain 'insularity', because they are locked in the British Foreign Office point of view. It often happens, for example, that conduct on the part of another government which at first seemed outrageous to us, becomes much more explicable when, and only when, we see other country's papers and learn, among other things, the misapprehensions under which the other party was labouring. Anxiety about matters of policy in his own day may increase the 'insularity' of the historian, even when he is dealing with the past; and those who are apprehensive of Germany in 1913 will be more easily satisfied with a hostile interpretation of Bismark. Fairly recent historians of Canning and Castlereagh[6] have perhaps been too near to Metternich to do him justice – they may have been, still, too closely involved with the issues of that period to range their history into form, save by the method of 'taking sides'. In this sense, history has sometimes to be regarded as 'action' as well as 'thought', and something of a man's attitude to the past may enter into the structure of his own contemporary world-view, and even of his policy. An historian of imagination may divine that the conduct of another government will prove to be explicable, before he knows what the explanation is; for the historian has to find a way of taking account of the evidence he does not possess. Even here, however, the position of the historian must be closely analogous to that of the statesman. Both Crowe's memorandum and Sanderson's counter-memorandum are extraordinarily historical in character. But Sanderson himself ceases to be historical in his final paragraph, except in the sense that his policy recommendations seem to depend on a way of conceiving the process of things in time.

Crowe's querulous attitude towards Germany reminds one of those colonial servants who used to see aspirations to freedom only as a nuisance to the administration. Because he is so definitely making out a case – accumulating all the things on one side of the question – he intensifies his insularity, giving a surprisingly crude version of the earlier history of Brandenburg and Prussia. In the years immediately after the First World War, scholars tended to praise the historical side of his memorandum; but this was a period when anti-German feeling was still strong, and the case still had to be made for the continuance of penal measures against Germany. The same scholars would have been less enthusiastic if they had known that the military historian, Spenser Wilkinson[7] – an anti-German agitator in the days when Lord Roberts[8] was breathing war, and a relation by marriage of Eyre Crowe[9] – had claimed the responsibility for the historical side of the memorandum. Even if historians today came down on the side of Crowe in the case of every diplomatic episode under discussion in the papers of 1907, Sanderson's view would have to be regarded as the more 'historical' because of its catholicity – because it widened the range of considerations involved. It was he, moreover, who, instead of merely speaking from one side of the Anglo-German controversy, tried to stand above the parties, surveying the problem from a higher altitude.

Crowe sees England and France as peace-loving, and does not remember that, until very recently, the historical case now made against Germany had been made, more unreservedly, and with equal plausibility, against France. He is so convinced of the righteousness of British diplomacy that he is uncomprehending when the Germans, after they have supported us, expect a *quid pro quo* – a bribe for merely doing what is just. He is stung by the German preference for what Sanderson calls 'a traditional view of negotiation' – the view that 'one of the most effective methods of gaining your point is to show how intensely disagreeable you can make yourself if you do not'.

Sanderson, in his criticisms of Crowe, follows a line that could almost be reduced to a mathematical formula. In his paragraph,[10] for example, he regards it as significant that M. Delcassé 'ignored Germany entirely when he commenced operations in Morocco'.

> The German Government and the German nation are extremely sensitive about being ignored or neglected in the discussion of important questions, and it is not surprising that on this occasion they should have been much exasperated.[11]

Sanderson does not agree that Bismarck deliberately cheated England at the inception of a German colonial policy. 'Prince Bismarck was personally opposed to German colonization.... Suddenly he found that the movement was too strong for him.'[12] In regard to Zanzibar 'the procedure of Germany was no doubt annoying to us, but we have no claim to a monopoly of acquisitions in Africa'. Again, Britain made to Germany certain concessions (which were to prove delusive) in regard to the future of the colonies of Portugal; and after this the Germans, though they drew away from Krüger,[13] still went on behaving in an unfriendly manner. 'But,' says Sanderson, 'their conduct towards us, though not particularly gracious, was perfectly correct.' In the case of one of their protests against our right of search, they were provoked by 'an act which if practiced on ourselves, would certainly have been denounced as intolerable'.[14] The underlying pattern – the mathematical formula behind Sanderson's criticisms – is to me astonishingly like that behind Harold Temperley's criticisms of Professor Pearce Higgins in 1927.[15] (Temperley refused to allow these criticisms to be identified with his name until he had finished the publication of the *British Documents*, and they appeared in the *Cambridge Historical Journal* for 1948, pp. 251–56).[16] It is a question partly of viewing Anglo-German relations from a higher altitude, partly of 'seeing ourselves as others see us'; and Sanderson writes: 'If the mere acquisition of territory were itself immoral, I conceive that the sins of Germany since 1871 are light compared with ours.'

Finally, surveying the whole matter with something like the presidential eye of the historian, he says:

> We have to remember that the [German] Empire took its present place among the Great Powers of Europe only 35 years ago.... It was inevitable that a nation flushed with success which had been obtained at the cost of great sacrifices, should be somewhat arrogant and over-eager, impatient to realize various long-suppressed aspirations, and to claim full recognition of its new position. The Government was, at the same time, suffering from the constant feeling of insecurity caused by the presence on the East and West of two powerful, jealous and discontented neighbours. It is not surprising that with the traditions of the Prussian monarchy behind it, it should have showed itself restless and scheming, and have had frequent resource to tortuous methods, which have not proved wholly successful.... But, as time goes on, her manner of pursuing these objects will probably be less open to exception.[17]

Sanderson's memorandum differs from Crowe's, in fact, because it assumes a more dynamic view of the time-process, and is prepared to start its diplomatic reflections on the basis that all Great Powers are naturally striving to expand.

In both memoranda the discussion of history slides so easily into the consideration of policy as to suggest that one's attitude to the past is hardly to be separated from the question of what one wants to do with the world. In his final paragraph, no. 28, on policy,[18] Sanderson concludes,

> Germany is a helpful, though somewhat exacting, friend...a tight and tenacious bargainer and a most disagreeable antagonist.
>
> It is, as a general rule, less inconvenient to take her at once into counsel, and to state frankly within what limits you can accept her views, than to have a claim for interference suddenly launched on you at some critical moment. It would be absurd of course to make her any concessions of importance except as a matter of bargain and in return for value received.... But I do not think it can be justly said that she is ungrateful for friendly support. It is at all events unwise to meet her with an attitude

> of pure obstruction... It would be a misfortune that she should be led to believe that in whatever direction she seeks to expand she will find the British lion in her path.[19]

The passages quoted in Desmond Williams's paper show that Crowe, while admitting that Germany might be merely using every legitimate opportunity for the expansion of her interests, thought that this might slide later into a conscious attempt at 'a general hegemony and maritime ascendancy'. His immediate recommendations were nearer to Sanderson's than might be imagined, (as he himself noted,)[20] and it was rather the mood and 'tendency' of his memorandum which was calculated to affect Britain's angle of approach to the whole question. On the whole, while Sanderson met the German problem by offering tips for negotiation, Crowe suggested the possibility that here was a new candidate for 'universal domination'.

It could still be held that the Sandersons are a danger in politics, that the more 'historical' attitude is not the appropriate one for the practical world, and that, in any case, the historian would always be too academic to be a good advisor. It might be argued that if our 17th century parliamentarians had had a catholic vision, and a truly historical view of *Magna Carta*, their knowledge would have paralysed them: though it must not be assumed that the alternative would have been despotism, for liberty might still come to the world even if men are reasonable, and (as the 17th century showed) it does not always come most happily through revolutionary means. Our judgement may depend once again on what we want to do with the world; and it is safest to assume that men act most wisely when they have fuller knowledge and conform to reason, however useful passion and narrow-mindedness may appear to have been at times. What we must say is: that if certain safeguards are overlooked, a statesman can fall into error through too great a capacity for understanding the other party, or too great a desire to understand. If this is his one aspiration, he had much better give himself to the trade of an historian. It can hardly be true in the absolute that the statesman can have too comprehensive an understanding; but the capacity to see the other party's point of view sympathetically may be disastrous to him unless his mind embraces wider considerations still.

Sanderson sees the nervousness of a new power or the effects of unskillful diplomacy, where Crowe sees malignancy and hostile purpose, or sees these as existing potentially – requiring to be taken into consideration straight away. Under the conditions of 1907 it might have been the case (though I am not clear that it was the case) that the man preoccupied with the prevention of war between England and Germany would lean to the side of Sanderson. Another man might argue, however, that at a certain stage, the object of preventing war must be superseded if necessary by the aim of improving one's position and prospects in the event of war actually breaking out. Once this latter object becomes the more immediate of the two, the balance is more likely to be tilted in favour of Crowe; but in general these two fundamental purposes must be entangled with one another in the operations of diplomacy. They underlie Western diplomacy in the period 1945–66.[21]

The real difference between Sanderson and Crowe is in the structure of their statesmanship, and the real issue between them lies here rather than in the presence or absence of 'British' forms of prejudice. *A priori*, the anti-Germanism of Crowe might conceivably be regarded as the result rather than the cause of the assessment that he made of the German problem. Granting that the prejudice came first it was in any case personal – not the necessary result of the outlook of the Englishman *qua* Englishman. For this reason, it is significant that Sir Edward Grey, in *Twenty-Five Years*,[22] confesses that he brought an anti-German bias with him when he took over the Foreign Secretaryship. He traced it back to his experiences at the Foreign Office in 1892-95. In this earlier period, he had been outraged by the very things that had offended Crowe; and even in 1910, he wrote to Sir E. Goschen[23] of that earlier period: 'It is true that the Germans did support us in Egypt, but I remember once at any rate a threat to withdraw that support if we did not clear the way for German railway concessions in Turkey.'[24] It seems to be recognised that our foreign policy was weak in 1892-95, and a certain Pharisaism may have continued to spoil our relations with Germany during the Foreign Secretaryship of Grey, so that, if that country was rough and tortuous in its diplomacy, we gave some provocation through our unimaginativeness. Perhaps we were more flexible and imaginative in our handling of France and Russia, who, at the beginning of the story, constituted our essential problem. I suspect that we were like a former colleague of mine who quarreled bitterly with his friends for offences so minute that they hardly knew they had done wrong, but could be wonderfully generous to an enemy (or even to a friend who had offended very seriously) since he was then alerted to the fact that a great call was being made on his charity.

Crowe's memorandum must be envisaged as 'action' as well as 'thought', and its manner is more comprehensible to me if it is regarded as propaganda rather than as an attempt to hammer out a foreign policy. While it is being considered in this latter aspect, however (i.e. as a series of reflections on the situation) attention ought to be called to what perhaps is its most remarkable feature. It treats of Germany in isolation, without surveying the whole diplomatic scene in the way that one expects, and often finds, in a document of this sort. In particular, it makes no attempt to measure the danger from Germany against the danger from Russia either at that time or in the more distant future; though Englishmen still feared the Russian bogey, too, and many of them regarded the Czardom as sinister and dark. The critics of Grey's policy were the people who were able to see both the German and the Russian danger at once and there were many of them in the Liberal Party down to 1914, when Morley[25] asked what would be done about Russia after German power had been destroyed.

The Foreign Office documents down to 1914 are tantalisingly elusive about the precise question of the Russian Bear. Perhaps there is a basic prejudice, deeper than all the other prejudices, which induces a man like Crowe to apply Sanderson's mode of treatment to Russia, but not to Germany, while others apply it to Germany, but not to Russia, and few seise on it as a 'science' that ought to be applied all round. If Crowe cites the instances of German violence and aggressiveness, Temperley, when answering similar allegations, points out that here was just where Russia was worse than Germany; so that Crowe had perhaps simply decided which of the two Leviathans he was ready to 'appease'. By widening his survey, however, he might at least have produced a qualifying formula, and recognised the immediate problem as being only an interim one: the problem of holding off Germany until Russia became the greater danger. On such a system, it would have been less easy for us to make the utter destruction of Germany our avowed policy at a later time. The truth, however, is that a certain profound fear of Russia does leak into the British Foreign Office documents in the years before 1914; and, so far as I know, this whole problem has not yet been adequately treated by historians, though it is a matter of crucial importance. Perhaps the justification of Sir Edward Grey and Sir Eyre Crowe depends on the validity of their notion that Germany and Russia might ally with one another against us, unless we forestalled this by joining with one of them.

So far as the 'national prejudice' against Germany is concerned, this itself, whether in Grey and Crowe or in the public at large, was hardly a first cause – it was the effect of experience, even if the experience was being misinterpreted. The anti-Germanism was itself being manufactured in this period, and this was partly no doubt by the action of government. Towards the close of the year 1910, Bethmann Hollweg[26] pressed for a discussion of this anti-Germanism, and of the government's part in it; and it is clear from our reply[27] that ministerial propaganda against Germany had been provoked by the necessity of reconciling Englishmen to the cost of the naval competition. We were wrong to imagine that the Germans had no right to build a big navy. It is always a pity when the requisite armaments or counter-armaments cannot be built up without alarming the populace against some particular neighbour – without enflaming people with the fear of a 'bogey' somewhere. The conclusion of the matter is, perhaps, that one cannot escape the necessity of educating democracy in the fundamentals of foreign relations. The Kaiser's government, by its naval policy, did the most to ensure that anti-Germanism would be inevitable and to forge that chain of necessity (if there was one) which was bound to give Crowe the victory.

For anti-Germanism had not been a British characteristic; and it would be folly to suppose that a British government could not have turned it into anti-Russianism. My own memory of scares about the Russian bogey before 1914 is at least as vivid as any memory that I have of the German danger. I was brought up on the view that English liberty went back to the Teutonic forests, and that our parliamentary system was due to the Germanic elements in our constitution. Germany was associated with federation and free cities, as well as with Bismarck; associated also with the freedom of the Reformation – it was the Latin countries that had the papacy, the inquisition, and the Napoleonic despotisms. I heard the industrialists of the West Riding boast about the way in which their educational developments followed German models. I knew their sons, who so often, down to 1914, were sent to Germany (instead of going to our universities) to complete their education. British Protestantism was still under the leadership of Germany, and, as one German theological professor succeeded another in the seat of authority, I heard non-conformity taunted for having German professors as its virtual popes. I remember how often our toys were marked 'Made in Germany', and a certain jealousy of German industrial success would appear spasmodically in the populace, in the way such things appear at the present day. Lloyd George's Insurance Act[28] was a further example of German influence. If Germans can trace back their divorce from the rest of Western Culture to the 19th century, and if the Germanic theory of the British constitution was being undermined in the 1890's, these phenomena related only to the thought which existed at higher levels. English and German historiography were closely related to a Burkeian historical legal-constitutional system that both countries felt themselves to have in common; and only in very recent decades have we left the older tradition, and moved over to the more definitely 'Whiggish' progressivist kind of history, previously more familiar in France and the United States. The German professors cut a great figure in London at the International Congress of Historians in 1913. And John Clapham,[29] whom I always remember as the most stolid example of a Britisher, though he declared his devotion to his country at the outbreak of war in 1914, wrote that he was only sorry that we were not fighting the Slavs instead of the Germans. A number of Cambridge dons, particularly Liberal ones, made printed protests against the outbreak of war with Germany. It is difficult to believe that anti-Germanism was inherent in the British, or that a government could not easily have produced anti-Russianism instead. If government had no choice in the matter, the 'necessity' which fixed the direction of its propaganda was one that arose out of the naval competition.

II The role of human decision

In our discussion of Desmond Williams's paper, or in our talks about decision-making, I seem to remember that for a moment we were presented with the picture of England sliding unconsciously into an attitude of hostility to Germany.

Historians are often criticised for reading the story backwards and imagining that Bismarck intended from the first the things that he finally achieved. They run together fifty of his successive adjustments to a changing situation, making these solidify into a mighty act of will; and therefore they tend to see policy as more purposeful and far-sighted than it really is. I believe, however, that these dangers are widely recognised in the profession, if they were not first discovered in the historian's own examination of the historical process. And perhaps the mistakes are really the optical illusions of abridged history, not consistent with the actual process of research (though imposing themselves unnaturally sometimes on the history that research students write). The historian, for his part must be on his guard against those who govern us; for these might be over-ready to disown responsibility, to blame that public opinion which they help to form, and to assert their own helplessness before the flood of events or the complications of their own machine. I have half a memory of Sir Edward Grey's account of a Foreign Secretary, confronted by the morning's mail, pushing everything possible on to other departments, and answering the things he had to answer – all the time watching the clock, because his heart was really with the birds. Grey would have been a poor Foreign Secretary if there was no more to the business than he was trying to suggest; and, if his account is to be taken at its face-value, this may help to explain why, in the 1920s men like Temperley regarded him as too much under the thumb of his permanent officials. Whether the country was merely drifting into hostility with Germany in a fit of absence of mind is ultimately a matter for the historian, who will presumably collect all the various kinds of evidence in the long run. But granted this, one might have to say, at the finish, that faulty diplomacy, or a faulty diplomatic machine, must have been responsible for slipping and sliding of such momentous significance. Crowe's memorandum suggests that even a government on this sort of inclined plane has moments when it can take stock, or take its bearings afresh.

There are signs that, in spite of his gentleness or gentility, Sir Edward Grey, as Desmond Williams says, 'was, in some respects, of steel-like character'. We have to reckon with the view (held by some people) that he impressed upon British policy an anti-Germanism that must have been anterior to the influence of Eyre Crowe. As soon as independent historians began to inquire into the story, they seised on the fact that Grey concealed the continuance of the Anglo-French military conversations from most of the Cabinet – concealed them even from Asquith[30] (though in the meantime Asquith had become Prime Minister) until 1911, when the revelation produced a certain amount of alarm. This seemed to show that Grey realised the serious implications of the continued Anglo-French talks, but that – no matter what the Cabinet might feel – he was grimly determined to his way. When cross-questioned by historians after the First World War, it was at this point that his answers were the most shuffling and unsatisfactory. The strongest argument for relentless historical process, overriding the will of men and even of Sir Edward Grey, is the suggestion that, even before Grey became Foreign Secretary, this process had already begun, and the *Entente* had been turning into something more like an alliance. I am not clear that there is any evidence for this, and there seems to be no convincing evidence for this, and there seems to be no convincing evidence in Gooch and Temperley; but there is evidence against it, for Grey on one occasion mentioned the difference between his policy and Lansdowne's,[31] and he seems to confirm the view that the difference lay exactly here. Writing to Goschen in 1910 he says: 'My impression is that Lord Salisbury and Lord Lansdowne always wanted friendship with Germany.[32] It is true that he added that he also wanted 'friendship' with Germany; but the context makes it clear that he was pointing to a contrast between himself and his predecessors. When the hostility with Germany came to mean more definite alignment with Russia – a power associated with despotism, anti-Semitism, and ruthless aggression – Grey met with serious resistance from members of his own party in the House of Commons, though the Conservative party supported him. His attitude to Germany, whether right or wrong, is more understandable than his attitude to Russia. Perhaps the real source of the anomaly was Francophilism.

On the other side it can be argued that Grey's strength lay in his character rather than his intellect – that he had determination but did not have a calculated policy. The historians who cross-questioned him on the continuance of the Anglo-French military conversations were puzzled by his muddle-headedness and obtuseness, and Temperley was not sure whether this was deliberate. If Grey had little use for far-sightedness, his mind would be more subject to immediate impressions, and to the mood produced by recent events. It is curious that in 1907 and 1910 (and in respect of the events of 1892–95) our complaint always related to the 'hostile' demeanor or the bad manners of the Germans. And it was for this reason, (we told them) that we could not regard the German naval expansion in the way we regarded that of the United States. The Germans in turn objected to our alignment with their enemies, and asked us to explain our hostility; but what was absent was a genuine conflict of interests of the kind which had recently made war with France or Russia seem inevitable. It was as though good manners mattered more than *Realpolitik*; and if *Realpolitik* had been in question, we might have seen that, so far as the responsible actions of government were concerned (and ignoring the cries of unreasonable parties that existed in every state) the one serious movement for a change in the territorial *status quo* in Europe in the years before 1914 was the Slavonic one – the Austrian aggressiveness being provoked by this threat (with Conrad[33] probably in the right when he said that, by postponing war for so long, the Hapsburgs were throwing away the last chance of winning it). At the same time some of the unfortunate German actions which had been responsible for earlier crises and for Grey's complaints, had been provoked by changes in the *status quo* for which France had been responsible. If you looked at the disposition of forces on the map, the main clashes of interests that seemed inescapable for Britain seemed still to be those with Russia; though if you were merely following your nose and taking a line of least resistance, you might feel that by supporting Russian ambitions in Europe you would be slackening the tension in Asia for a time. On this whole view of the matter, if Grey had a policy it might have been a policy of mere drift; but we need not imagine that such a policy was inevitable, since it might have represented a weakness in Grey's Foreign Secretaryship.

In the years after 1919, historians tended to opt for the second of these alternatives, and to say that Grey tended to drift. They disparaged his intelligence, but they were capable of doing this too lightly and flippantly. It seems clearer that Grey set out in a remarkable manner to play the mediator in the years just before 1914; and he must have had some success in this – he was even happy to see that Germany showed a willingness to cooperate. His alignment with France and Russia interfered with this policy, so that we are left with an impression of what he might have done if he had been independent of the competing groups. He was terrified of 'isolation', however, haunted on occasion by the dread of a Russo-German alliance, perhaps too much governed by fear, and unable to see the strength of Britain's diplomatic position. He came to be afraid of being considered an unwilling ally, and this seems to have been a significant factor in July 1914[34] by which time it had become an obsession with other powers too. Yet, treaties with Russia and Italy in 1915 would suggest that it was the unwilling ally, the exacting government that got the best price. In the total result, Grey at least was not so strong a leader as a country could have desired.

I cannot resist the final feeling that in reality there was a strong will at work in British foreign policy in the age of Sir Edward Grey, resolutely impressing itself on events, and definitely the leader, not the led. It is not exactly represented in Crowe's memorandum of 1907, but the memorandum may serve to typify it, as I believe Desmond Williams intended it to do. It has been (and I suppose it still is) characteristic of our governing circle that a firm group of people form a combination, united by a common outlook, a body of unquestioned assumptions; and it seems to me that if you do not accept their basis of reference, they can nullify your influence – you are an 'outsider,' you soon learn that you are not in the inner ring. If you are in the circle, you churn everything into your system, and – in the realm of foreign policy – an ambassador will learn that it is useless to fight against the system, or he will be regarded as a recalcitrant whose dispatches are ruthlessly discounted, or his reports will be turned somehow to the service of the appointed cause. I have often been surprised at the efficiency with which bright and independent young men are absorbed into the system.

From this point of view, the relations of Sir Eyre Crowe, not only with diplomatists, but also more particularly with the military – and indeed with Spenser Wilkinson and Wilkinson's circle – are deserving of great attention, as Desmond Williams suggests. A research student of mine who was working on the military and naval attaches in the years before 1914 was struck by the significance of the military members of the group. He came to the conclusion that Crowe was ready to pooh pooh the reports of ambassadors and to prefer the views even of military attaches, in a way curiously analogous to that of Kaiser Wilhelm II. If mere prejudice lay behind the policy of this circle, it was not 'national,' and, as Desmond Williams suggests, it was rather peculiar to the group – with the diplomats concerned, it may have been Francophilism. As Williams points out, this does not necessarily mean that their policy was a wrong one; and a country is perhaps fortunate if it enjoys a strong, self-conscious leadership, so that it is not left to drift, or locked in historical processes. A country is perhaps more fortunate still if the minister keeps command over its policy.

At any rate, it would seem that while many talk of drifting, and are content to drift, a handful of resolute people may be firmly subjugating history to their will.

Notes

* Presented on 15 July 1960 (eds).

1. Delivered to the British Committee on the Theory of International Relations on January 9, 1960. CUL, File 329 (eds).
2. Presented at the same session (eds).
3. G. P. Gooch and H. Temperley eds *British Documents on the Origins of the War, 1898–1914* (London, 1926), III, pp. 397–420.
4. Ibid., 420–31. Butterfield is here referring to Thomas Henry, Baron Sanderson (1841–1923) civil servant and permanent undersecretary 1894–1906. He was an able administrator noted for his gifted composition of official dispatches. See B. Vegezzi, *The British Committee on the Theory of International Politics, 1954–1985* (Milan, 2005), p. 324 (eds).
5. Gooch and Temperley, III, 420.
6. Butterfield had in mind: H. Temperley, *The Foreign Policy of Canning 1822–1827* (London, 1925) and C. K. Webster, *The Foreign Policy of Castlereagh 1815–1822: Britain and the European Alliance* (London, 1925) (eds).
7. Henry Spencer Wilkinson (1853–1937) military historian and journalist, Chichele Professor of Military History, All Souls, Oxford 1909–23. He was a moving spirit behind the foundation of the Navy League.-(eds).
8. Frederick Sleigh Roberts, Earl of Waterford (1832–1914). Field marshal and Commander in Chief 1900–1905. Fervent advocate of National Service. (eds).
9. Sir Eyre Alexander Wichart Crowe (1864–1925). Diplomat Asst. Undersecretary for foreign affairs 1912, and permanent secretary 1920–25. He was an ardent advocate of close Anglo French relations. Henry Wilkinson was married to Crowe's sister Victoria. (eds).
10. Gooch and Temperley, III, 421.
11. Ibid., Théophile Delcassé (1852–1923) French statesman and foreign minister (1898–1905, 1914–15) who largely shaped France's alliance system before World War I (eds).
12. Ibid., 422. Gooch and Temperley, III, 422.
13. Stephanus Paulus Krüger (1825–1904). President of the South African Republic from 1883, and renowned for his resolute leadership during the Boer War, 1899–1902 (eds).

14. Gooch and Temperley, III, 426.
15. The response to a lecture entitled 'The Law of Nations and the War', and later published as part of a collection, *Studies in International Law and Relations* (London, 1928) (eds).
16. 'Professor Temperley on the Origins of the war of 1914', *Cambridge Historical Journal*, vol. IX (1948) pp. 251–6 (eds).
17. Gooch and Temperley, III, 426–7.
18. Ibid., III, 430.
19. Ibid., III, 430–1.
20. Ibid., III, 431.
21. And, one might say, from that period to the present, if not beyond (eds).
22. Viscount Grey of Falloden, *Twenty-Five Years, 1892–1916* (London, 1926) (eds).
23. Sir William Edward Goschen (1847–1924) British ambassador at Berlin 1908–1914. (eds).
24. Gooch and Temperley, VI, 538.
25. John Viscount Morley of Blackburn (1838–1923) Statesman and writer, advocate of social, land and tax reforms, with a pacifist stance on foreign and imperial policy. (eds.)
26. German Chancellor, 1909–17. (eds).
27. Gooch and Temperley, VI, 558.
28. Passed in 1911 (eds).
29. Sir John Harold Chapham (1873–1946). Professor of Economic History, University of London 1928–1938. Works include *An Economic History of Modern Britain* 3 vols (eds).
30. Herbert Henry Asquith (Earl of Oxford and Asquith 1852–1928). Liberal Imperialist during the Boer War, Chancellor of the Exchequer 1905–08; prime minister 1908–16. He was a close supporter of Sir Edward Grey, a strong proponent of the entente with France, and the creation of security measures against the growing German naval challenge (eds).
31. Henry Charles Keith, fifth Marquess of Lansdowne (1845–1927). Viceroy of India 1888–94, secretary of state for war 1895–1900 and secretary of state for foreign affairs 1900–1905. Unlike Grey, he favored a policy of reconciliation with Germany and in 1917 proposed a 'peace of accommodation' with the Central Powers which resulted in accusations of disloyalty to the allied cause and led to his repudiation by the conservative party (eds).
32. Gooch and Temperley, VI, 538.
33. Franz Conrad von Hotzendorf (1852–1925) Field marshal and head of the Austro-Hungarian General Staff 1906–17. An aggressive militarist, he actively supported Austria's declaration of war against Serbia in 1914 (eds).
34. Gooch and Temperley, XI, 81–2. cf G.P. Gooch, *Studies in Diplomacy and Statecraft* (New York, 1942) pp. 87–107 (eds).

6
Sir Edward Grey in July 1914*

Concerning the Germans, the Cambridge professor, Alfred Marshall, said in 1914 that we 'must fight them with all our might' but it was 'our duty to respect them'. After Britain's entry into the war against them, he described them in *The Times* as 'conscientious' and 'trusty'; and, though he insisted that they were to be feared, he took the line that they were 'not to be vilified'.[1] A number of academic Liberals in those days would have agreed with G. M. Trevelyan, who wrote to the *Daily News* on 3 August that 'everyone should do something to demonstrate his or her antagonism to the participation of England in the European crime'.[2] I have seen a letter in which the later Sir John Clapham of King's College wrote on 2 August to Professor Whitney angrily protesting against the idea of fighting alongside the unspeakable Russians against Germany. 'I am a Teuton', he said, 'not a Slav.'[3]

The Liberal party, which then held office, had been divided on the issue of foreign policy ever since the South African War. It has been asserted with some plausibility that in July 1914 it was 'the old pro-Boer contingent' who, at the National Liberal Club were 'in the main isolationists'. G. P. Gooch tells us that the South African War was the 'turning-point' in his life. It drew his interest away from social questions and made him feel that his mission was to be in the realm of foreign policy. He became a critic of Sir Edward Grey, and in 1917 (amid the fever of a war in which so many historians lost their heads) he published what is probably the most permanently satisfying paragraph ever written on the subject of Britain's pre-war diplomacy.[4] According to the *Journals* of Lord Esher (III, 174), even two days before the British government declared war, 'twelve members of the Cabinet were against giving any definite assurance' to France and Russia. In other words, they were still declining to follow the lead of Asquith, Grey, and the service ministers. When we discuss the origins of the First World War, we ought to bear in mind the fact that we do not quite know what England would have done if the Germans had not invaded Belgium.

There are certain pitfalls in the study of the origin of any war, and, as they have affected the inquiry into the causes of the war of 1914, a word about them may be useful. And first of all there is the difficulty of imagining ourselves on the other side of the month of August – recapturing the state of mind of even the majority of the British cabinet before it was known which way the dice were going to fall. Secondly, as Professor Temperley once pointed out, we may be too apt to assume that the plans and ambitions of the military department really represent the foreign policy of a government.[5] In particular, there are reasons why plans brought forward after the outbreak of war should not be given a retrospective validity as an historical source – not assumed to represent the considerations on which a government acted when it was entering the war. For this reason one should be careful in the inferences one makes from the material supplied in Fritz Fischer's *Griff nach der Weltmacht, Kriegszielpolitik des kaiserlichen Deutschland 1914–18*. On similar grounds, the evidence drawn from Pan-Germanist propaganda ought, initially at least, to be compared with parallel evidence drawn from the Pan-Slav movement or from militaristic agitators in Great Britain – not with the responsible policy of the British government. There may be a kind of *demi-monde* composed of people at the sub-governmental level – people who behave somewhat disreputably precisely because they are sheltered from public responsibility. If we have evidence of these from one country but not from another, here is a point where it might be well to take the evidence with the tongue in one's cheek. At least it is important to be austere in distinguishing the decisions that represent the official policy of the government.

Thirdly, it is difficult to avoid the danger of seeing another country's foreign policy through the eyes of one's own Foreign Office. Dr Gooch was making a significant point when he said on one occasion that the first task is to learn what policy 'looks like to its makers and friends'.[6] A measure which, to the outsider, looks wickedly uncalled-for, sometimes becomes more understandable when one sees how the decision was arrived at, or discovers the predicament that provoked it, or learns the considerations on which it was based. For this reason, the historian needs not only the official correspondence – the dispatches between a Foreign Office and its embassies abroad – but also the policy-making documents and the private letters that relate to policy decisions. Even so, there is always need for criticism and careful balancing because sometimes we possess the crucial evidence more completely for one country than for another. Moreover, one man may put on to paper every wild idea that comes into his head, while another can move unrelentingly to his object, covering himself at every turn and giving as little as possible away.

To see policy as it is envisaged by the policy-maker is particularly important in view of the remarkable operation of fear in international affairs. Here is one of the most insidious of the pitfalls in the study of the origin of wars; especially as fear so often produces a strange obliquity of conduct, and nations – even more than individuals – tend to feel that they must not show their fear to the outside world. In the Second World War the British were sometimes haunted by the fear that the Russians were going to desert them and to patch up a kind of compromise peace with Hitler. Before 1914, there were times when the most terrifying thought was not that the conflicts of Europe were leading to inevitable war, but that the continental powers might make up their quarrels and combine against the British Empire. The fears that operate on governments may be fantastically implausible; and the state that dreads too many alternative conjunctures may look for too many guarantees against hypothetical possibilities, or else it may be too paralysed to have any diplomacy at all. Historians, when they are discussing foreign policy, tend greatly to underestimate the operation of fear, in the case of every country but their own, because it is so difficult to put oneself in the other party's place.

Finally, so long as people argue that an historian is a 'pro-German' when he adopts a certain interpretation of the origin of a war – so long as they say that a given historical interpretation is inadmissible because it would encourage too relaxed a policy towards Germany at the present day – so long as they are governed by this kind of preoccupation – they cannot claim to have achieved at the same time a detached historical judgement on the events of July 1914.

II

The serious hardening of English policy towards Germany had occurred after the defeat of Russia by Japan in 1904–1905 – a catastrophe which left Germany for a time with a formidable position on the map of European forces. It happened also that in 1905 British policy came under the direction of Sir Edward Grey, a man who reveals in his memoirs that in the 1890's he had acquired at the Foreign Office what some of his colleagues called anti-German prejudices.[7] On 1 January 1907, Sir Eyre Crowe, then Senior Clerk of the Foreign Office, produced his famous Memorandum, which suggested that Germany ought to be treated as a power making a drive for political hegemony.[8] Crowe called for more solid and continuous British co-operation with powers like France and Russia which were ready to counter the German danger. His brother-in-law, Spenser Wilkinson, claimed part of the credit for this Memorandum, but he was also involved in the anti-German agitation under Lord Roberts, while the military conversations with France had brought Crowe himself into closer collaboration with important army officers.[9] Sanderson, who had recently retired from the post of Permanent Under-Secretary at the Foreign Office, produced a Counter-Memorandum containing a more historical explanation of German policy and criticising Britain's unimaginativeness as partly responsible for some of the difficulties that had arisen. But it was decided to take no notice of Sanderson.[10] It has recently been shown that, chiefly through the activity of Francis Bertie and the later Lord Hardinge, there had been occurring within the Foreign Office itself in this period a shifting of personnel which gave the predominance to the anti-German party.[11]

Before the crisis of July 1914 the situation had altered and the distribution of power had changed. Poincaré is significant for the tremendous intensification of the *Entente* from 1912 and he is identified with the rise of nationalist feeling in France itself.[12] Now the armaments of France and Russia began to seem formidable and disconcerting. Poincaré's achievement possibly made it more difficult henceforward to prevent the outbreak of a war, but secured that *if* war actually occurred, the *Entente* would have a better chance of winning it. The Belgians reproached Great Britain for joining one of the continental *blocs*; they resented the Franco-British military conversations which envisaged the defence of their neutrality – they feared that they might be invaded by powers which were determined to defend them against their will.[13] Their ambassadors in Paris, Berlin, and St Petersburg described the mounting armaments and increasing aggressiveness of the *Entente*, which they now regarded as constituting the threat to peace.[14] The two great *blocs* confronted one another on the Continent, but in certain respects Sir Edward Grey rose above them during the Balkan wars of 1912–13, co-operating with Germany in the cause of peace. This very fact gave rise to a certain amount of jealousy in Russia, where the government sought more definite commitments from Great Britain, playing on Grey's fears – hinting to him, for example, that Britain was a power on which the *Entente* could not rely.

The state of the alliance-systems themselves is illustrated in the famous episode of Liman von Sanders, which, at the end of 1913 almost provided what somebody has called the dress-rehearsal for the final crisis. This German officer was appointed to lead a military mission to Turkey and the Turks gave him the executive command of an army-corps in Constantinople. In spite of the indignation of the court of St Petersburg, the Germans refused to withdraw the man, saying that such a step would damage their prestige in Turkey. The Russians were already jealous of what Sir Arthur Nicolson called our 'intimacy' with Germany; and, in order to reassure them, Grey allowed himself to be induced to agree to a collective protest in Constantinople, one that was couched in rather extravagant terms, for which the French seem to have been responsible. It transpired, however, that without the knowledge of either Grey himself or some leading men in the Foreign Office, a British admiral, Limpus, had for some time been holding a parallel position in Constantinople, where he had the command of the Turkish fleet. Indeed, it turned out that we had 'just secured' what our ambassador in that city called 'a monopoly of dock-yards and arsenals' in Turkey for a period of thirty years.[15] And, curiously enough, the British ambassador himself took the line that, *if* the Admiral's powers were to be depleted, his own country's prestige in Turkey would be injured too.[16] A lucky chance enabled Grey to withdraw from the extravagant measure which had been in contemplation; whereupon the Russian Foreign Secretary became more urgent and importunate than ever before.[17] In repeated interviews with the British ambassador, he reiterated that he would regard the episode as the test of the *Entente*.[18] He insisted that the Triple *Entente* was stronger than the Triple Alliance; but in his view it could only prevail if Germany were made to see that it would have to face the British fleet. When asked if he meant to drive the issue to an absolute extreme, he denied that this was his intention; but it was his main thesis that the extreme would never be reached if the British really supported Russia, for in the face of this combination the Germans would simply give way.[19] The British ambassador in St Petersburg reported on successive occasions that, if we failed to collaborate, we must expect a future change in the attitude of Russia to Great Britain. What the Russian Foreign Secretary demanded now was a formidable joint *démarche* in Berlin; but Grey had the feeling that such a hostile and militant approach would rather make it impossible for the German government to give way. It turned out that Grey had been right, for the Germans responded to more reasonable treatment. His comment was:

> I don't believe the thing is worth all the fuss that Sazonow [the Russian Foreign Minister] makes about it; but as long as he does make a fuss it will be important and very embarrassing to us; for we can't turn our back upon Russia.[20]

From this moment, a stronger attempt was made to secure that Britain should be committed more definitely to Russia.

It would seem likely that, throughout this episode, the Russians were pursuing the technique that was customary when an attempt was being made to induce the British Foreign Secretary to commit himself more deeply with the *Entente*. Nothing made Grey more nervous than a suggestion from France or Russia that Britain was not sufficiently reliable to be valued as an ally or friend; and the Russians were in an aggressive mood at this time. When the Liman von Sanders controversy was practically over, their Council discussed the question of the country's readiness for war and secured an opinion from the Chiefs of Staff. The result was interesting in view of the prevailing German idea at the time (and the notion accepted since) that Russia needed a year or two more for preparation. It was agreed that Russia could face Germany already, provided only that she could be sure of the British fleet. On the other hand, it was decided not to attack the Straits directly at the moment (though the object was greatly desired). Such a policy would be bound to provoke a general war, and Russia, if she had to fight Germany, could not afford to have part of her army engaged with the Turks farther east. In April 1914, a Russian historian, writing to the *Preussische Jahrbücher* to explain his country's hostility to Germany said: 'It is now clear to the Russians: if things stay as they are, the way to Constantinople must be through Berlin.'[21] The agreement of 1915 in which the Allies conceded

Constantinople to Russia shows that the Russians were in earnest, and, after 1919, the Russian historian Pokrovsky revealed how anxious his country had been to secure that Turkey should be an enemy (and not an ally) during the war, the desire to gain Constantinople being stronger than the wish to strengthen the *Entente*.[22]

Supposing, in the period before the war, one had asked where there existed a really aggressive movement in Europe – a movement avowedly seeking a radical change in the existing *status quo*, if necessary by violent means – one would have had to point to the Serbo-Russian activity in the Balkans, which was known to involve the actual break-up of the Austro-Hungarian Empire. In relation to this the Central Powers were on the defensive, and Grey himself regarded Austria-Hungary as weak, though the Pan-Serb agitation was bound to provoke counter-measures in Vienna. We, today, can speak more assuredly than contemporaries were able to do about the reality of Russian designs on the Straits and Constantinople. Furthermore, by July 1914, the British were so aroused by Russia's offences in Persia, that Sir Arthur Nicolson, though he had greatly desired the connexion with St Petersburg, wrote to the ambassador in this city:

> It seems to me that our relations with Russia are now approaching a point when we shall have to make up our minds as to whether we should become sreally intimate and personal friends, or else diverge into another path.

He talked about 'those apprehensions which unfortunately do exist here in regard to Russian designs'.[23] On the very day of the Serajevo assassination – but before he had heard the news – the Russian ambassador had already written that his country would have to change its conduct, and that Germany was now gaining ground in official circles in London.[24]

In his Memorandum of 1 January 1907, Crowe argued that Great Britain had made an agreement with France because there had been disputes between the two powers which required amicable settlement. He maintained that there was no need for an agreement with Germany because 'there are no questions of any importance now at issue between the two countries'.[25] The point is one that we meet at times elsewhere; in the memoirs of another leading personage at the Foreign Office, Lord Hardinge, we read on one or two occasions that there are no issues between England and Germany.[26] In the case of Russia, if we accept Crowe's argument, there was a particular call for the kind of convention that was concluded in 1907. Here, there were not merely contingent fears for the future; there was a precisely locatable area of conflict, and the conflict was fundamental, for it involved the threat to India. When the *Entente* with Russia was attacked, Grey would defend himself by saying that the only alternatives to this policy were either a passive surrender to Russian aggression or a big forward movement in Asia to counteract it.

One of the remarkable features of the story, as can be seen from both the papers of the time and the justificatory memoirs written later, is this idea of 'containing' the aggressor in Asia by the device of making friends with him. One continually meets the view that Russia will hold her hand in Asia because she is now grouped with Britain – the British are on her side. In a certain sense Grey's critics were right, because, as we have seen, the Russian aggressions in Persia caused increasing difficulties down to the very month of July 1914. On the other hand, Grey himself, both before the war and after it, maintained that, however unsatisfactory the conduct of Russia was, it would have been worse if Britain had not concluded the agreement of 1907. It has been said that, roughly speaking, there was an alternation in Russian policy – now a great desire for a forward move in Asia, now an interest rather in aggrandisement in south-east Europe. In so far as this was true, the Anglo-Russian *Entente* would seem to imply that, in return for peace in Asia, we would support Russia in Europe, and Russian aggressiveness would tend to be diverted to the Balkans. It would appear that by July 1914 the Foreign Office itself had begun to wonder whether its policy was really producing a sufficient check on Russian movements in Persia.

By this time, as we have seen, Britain and Germany had come into closer co-operation with one another, while Grey himself was prepared to try to act as a bridge between the opposing continental groups. On 24 June, four days before the assassination at Serajevo, he received a message from the German Chancellor, asking that in the case of future Balkan emergencies, Britain and Germany should continue to collaborate, taking care to keep in close touch with one another. Grey was determined not to be parted from the *Entente*, and he made it clear even to the Germans that, though Great Britain, France, and Russia 'were not bound by engagement as allies, we did from time to time talk as intimately as allies'.[27] Indeed the Germans were aware of the fact that there was now question of a naval convention with Russia. On the eve of the crime of Serajevo, however, Grey was not by any means the blind victim of anti-German prejudice and he was well aware of the possible danger from the Slavs.

III

If the Russians and the Serbs were the source of the really aggressive movement before the war of 1914, the point of imminent danger, even before the assassination of the Archduke Franz Ferdinand, lay in the tendency of Austria-Hungary to play with the idea of a preventive war. I personally think that, when an empire finds its doom at hand, and sees no salvation save in the wild gamble of a war, it ought to accept its fate with quiet resignation, regarding it as the decree of Providence. But I do not know how many empires would be ready to adopt this view; and Englishmen, at least, ought to understand the problem that Austria-Hungary had to face in 1914; for here was an extreme case of what I should call the 'Suez' predicament. The declining power says, 'We have suffered too many kicks; we won't go under without a fight'; and it expects the rest of the world to stand by, just holding the ring, while it has its own localised conflict with the equivalent of Nasser. The statesmen of Vienna had been coming over to the idea of a preventive war in any case; and, when the heir-apparent was assassinated on 28 June, most of them determined that the answer should be a sudden attack on Serbia. They were persuaded, however, that it would be better to send an ultimatum first, though most of them were resolved that this should be so severe as to be unacceptable. Professor Temperley wrote later.

> Austria-Hungary [in 1914], I feel sure, acted from fear and a sense of insecurity. The longer the [Pan-Serb] agitation went on – the more certain her ultimate dissolution. Unless she acted, her existence was endangered. And it was on the law of self-preservation that she acted. She had determined so to act in the near future – and [had] drawn up a memo – which she had already decided to forward to Germany asking for support [,] *before* the assassination.

Temperley had devoted many years to the Serbs, who had become in a certain sense his chosen people; yet he seems to have been troubled on occasion by the facile condemnation of Austria-Hungary; and he wrote:

> I shall not be suspected of undue prejudice ag[ain]st Servia – but the Ballplatz – at any rate after Aehrenthal – ... acted in a spirit of deadly fear.[28]

The court of Vienna had to secure the consent of its ally, Germany, and this was actually granted by the Kaiser and the Chancellor on 5 July. It was most emphatically the decision of the civilian government in Berlin; and it was the kind of decision that might have been taken by any civil government, determined not to go under, not to wait, as a submissive victim, for a threat to materialise. The result was the famous 'blank cheque' presented by Germany to Austria-Hungary – the promise of unlimited support whatever decision the court of Vienna might take, and whatever the consequences of it might be. This was the step which at a later date drew upon Germany so much condemnation.

It is important to recapture the mood of the time and this can be admirably illustrated by a glance at some British reactions to the crisis. In the House of Lords, on 30 June, Lord Lansdowne said: 'No reprobation which we can express will presumably produce any effect on the miscreants.' A letter in the correspondence column of *The Times* on 2 July suggested that, on the following Sunday, the tune 'Austria' should be sung in all churches. I remember, as a schoolboy, seeing the *John Bull*poster, 'To hell with Servia', and the facsimile in the same paper of a document that pretended to show the complicity of the Serbian government.[29] Even so late as 17 July, the *Westminster Gazette* declared that the Austro-Hungarian government 'cannot be expected to remain inactive'. As late as 27 July, Sir Edward Grey himself was prepared to say that 'if [Austria-Hungary] could make war on Servia and at the same time satisfy Russia, well and good'.[30] On 3 August, the *Manchester Guardian* expressed the sentiment: 'If one could tow Servia to the edge of the ocean and swamp it, the atmosphere of Europe would be cleared.' Whether one looks at the British cabinet or the Liberal party or the London press, one is surprised to see how much unwillingness there was to follow Sir Edward Grey even at a late date when so many new factors and considerations operated in his favour. The German ambassador of those days, Lichnowsky, has sometimes been used as a witness against his own government, but I wonder whether he was correct in one of his main ideas: namely, his view that an Austro-Hungarian attack on Serbia would swing British public opinion to the side of Serbia. Sir Harold Nicolson wrote, in his life of his father: 'It was realised, of course, both in Berlin and London, that the Austrian government would be obliged to make some demonstration.' He adds that neither in Downing Street nor in the Wilhelmstrasse 'did there exist the slightest sympathy for Serbia. They were both anxious to see that country brought to book'. Nicolson makes it clear that he is envisaging what he calls 'a punitive expedition' – a localised conflict – 'merely a frontier question between Austria and Serbia'. He criticises the well-known American historian, Professor Fay, who, he says, goes wrong because he fails to understand how at that time 'everybody expected Austria to take strong action'.[31]

This illustrative material is relevant because it recovers for us something of the mood, something of the context, in which the *British Documents on the Origins of the War* ought to be interpreted. The relevant section of that work is Volume XI, but less than a quarter of this is devoted to the first three-and-a-half weeks after the assassination (that is to say, to the period down to 24 July) while the other three-quarters is concerned with just the final twelve days of the crisis. For a period of nearly a month after the assassination, we have in the volume only nine dispatches of Sir Edward Grey to any part of Europe, and two of these are nothing more than straight letters of condolence. (According to the editors of *Documents Diplomatiques Français* there is no dispatch on the subject of Austro-Serbian relations between the French ambassador in London and his government in Paris from 8 to 22 July.)[32] Neither Winston Churchill nor Lloyd George could remember the matter being brought before the British cabinet till 24 July.[33] Even when the Austro-Serbian crisis is mentioned in dispatches by Grey in the early part of the month, it is entangled with another diplomatic problem which the Foreign Secretary seemed to regard as the harder nut to crack. This was the need to secure that the proposal of an Anglo-Russian Naval Convention should not have a disastrous effect on relations with Germany. On 6 July, a quite separate Balkan issue was under discussion, and Sir Arthur Nicolson, the Permanent Under-Secretary at the Foreign Office, declared that apart from this issue – the issue of Albania – 'we have no very urgent and pressing question to preoccupy us in the rest of Europe'.[34] And in the letter (already mentioned) in which Nicolson on 14 July sees Britain and Russia almost at the parting of the ways, there is no reference to the Austro-Serbian problem, unless it is in a part of the document which Gooch and Temperley left unpublished.[35] Lloyd George tells us that in the weeks after the assassination 'our Foreign Office preserved its ordinary tranquillity of demeanour, and thought it unnecessary to sound an alarm even in the Cabinet-Chamber'.[36]

IV

Yet we should be mistaken if we were to infer that Sir Edward Grey was too ignorant or too unimaginative to realise the significance of the Serajevo crisis. On the contrary, he was the one statesman in Europe who received a hint of what the Central Powers had in mind. On 6 July, the very day after the Kaiser had given the 'blank cheque' to Austria, the German ambassador came to see Grey and said that 'he knew for a fact . . . that the Austrians intended to do something, and it was not impossible that they would take military action against Servia'. And, according to the very next sentence of the report which he himself made of this conversation, Grey's reaction to this statement must be at least a matter for surprise to us. The sentence runs: 'I said, surely they did not think of taking any territory.' And the German ambassador gave him the assurance that nothing like that was going to happen.

The ambassador then went on to say (as he had been saying before the assassination) that his country was frightened of both the military power and the obvious hostility of Russia. Germany, he continued, felt that, if she tried to hold Austria-Hungary back, her ally would accuse her of perpetually refusing to stand by her. Therefore, would Grey use his influence in order to mitigate feeling in St Petersburg? The ambassador repeated his point. 'There was some feeling in Germany . . . ', he said, 'that trouble was bound to come and that therefore it would be better not to restrain Austria, and let the trouble come now rather than later.' It is difficult to see how Grey could have been given a better hint of what was in store, and even of the considerations on which the German government had decided to present Austria-Hungary with the 'blank cheque'. Lichnowsky asked Grey whether he agreed that, in spite of all this, the two alliance-groups, which had recently come closer to one another, should be prevented from drifting apart. Grey did agree. He said he was disturbed by the news, but he gives no sign of protest. He did not leap out in anger against the suggestion of Austrian action or the German intention of doing nothing to stop it. Finally, he said that he was glad that the ambassador had spoken to him, and he describes himself as promising that 'if trouble did come, I would use all the influence I could to mitigate difficulties and smooth them away, and if the clouds arose to prevent the storm from breaking'.[37]

Two days later, he was keeping his promise. He telegraphed to Paris that he feared the Austro-Hungarian government would be forced by public opinion to make some *démarche* against Serbia. He said that 'in such an event, we must do all we could to encourage patience in St Petersburg'.[38] The French ambassador in London reports on the same day a conversation in which Grey spoke to him in very much the same manner. Forthwith, the Foreign Secretary began the work of placating the Russian ambassador. Again, he explained his fear that, owing to the force of public opinion, the Austrian government 'might be swept off their feet'. 'When the ambassador replied that 'he hoped that Germany would restrain Austria', Grey turned this obviously awkward question, and he never seems to have answered it. He said that the Germans were frightened at the hostility of Russia – doubly frightened because there had recently been a leakage of news about a possible Anglo-Russian Naval Convention. By certain prevarications he had attempted to deceive the Germans about this matter, but they had delicately allowed him to see that they were not deceived, and he was particularly anxious not to alienate them. In this very interview, in which he had broached the idea of Austrian action against Serbia, he induced the Russian ambassador to agree that his court 'should do all in their power to reassure Germany and convince her that no *coup* was being prepared [against her]'. Most curious of all, in the same conversation, the ambassador put to Grey a further significant question. He asked what there was to give colour to a serious Austro-Hungarian *démarche* against Serbia. On this point, of course, Grey had no information – he could only offer a conjecture; and, in doing so, he revealed the kind of thing which, at this date, he imagined would justify Austro-Hungarian action against Serbia. Never throughout this stage in the proceedings did Grey suggest that it might be necessary for Austria-Hungary to demonstrate the complicity of the Serbian government in the assassination. For the moment he suggested that it might be possible to establish a charge of negligence against that government, if, for example, it turned out that the bombs had been made in Belgrade. All this is Grey's own account of the interview, drawn up for the British ambassador at St Petersburg, and is therefore a good indication of the line he intended to take with Russia. Benckendorff, the Russian ambassador in London, drew up a long and serious report of the interview; and, according to him, Grey spoke in a grave and agitated manner, repeatedly saying that the more Russia menaced the Germans or gained the advantage in military power, the more closely the Germans would have to cling to Austria.[39]

These interviews occurred on 8 July, and on the 9th Grey held the further conversation that he had promised to have with the German ambassador. He now said that

> If Austrian action with regard to Servia [were] kept with certain bounds, it would of course be comparatively easy to encourage patience at St Petersburg.... There were some things that Austria might do that would make the Russian Government say that the Slav feeling in Russia was so strong that they must send an ultimatum or something of that sort.

But, even after this, he made a most important remark, and it is significant that in his own account of the interview, he recorded a promise to which he attached no conditions, no hint of reservation. He said: 'that I would continue the same policy as I had pursued through the Balkan crisis, and do my utmost to prevent the outbreak of war between the Great Powers'.

In a diplomatic document of this kind (and particularly at such a conjuncture as this) it is significant that he does not say 'prevent the outbreak of war' but 'prevent the outbreak of war between the Great Powers' – he will do his utmost to prevent a general war. And he adds: 'The greater the risk of war, the more closely would I adhere to that policy.'[40] It is relevant to note that in his *Twenty-Five Years* (I, 309) he gives away the fact that he had contemplated the matter 'as one to be dealt with by Austria alone'.

Even on 15 July the German ambassador reported to his Foreign Office:

> Sir Edward Grey said that everything would depend on the kind of intervention *(Eingriffe)* that might take place; in no case ought there to be question of a diminution of Serbian territory. He has also, as has been reported, been striving to influence St Petersburg in favour of Austria's claims. But if there should be a violent commotion in Russia as a result of Austria's military measures, he would not be at all in a position to control that country's action, and, in view of the feeling of dislike toward England which prevails at the moment in Russia ... he will have to pay regard to Russian sensibilities. In any event I am sure that the Minister will do his utmost to hold Russia back in the event of an Austro-Serbian conflict. But I do not believe he is in a position to speak the decisive word there in the way that he might do in Paris.[41]

Lichnowsky, who produced this statement, had already expressed doubts about the policy of the courts of Berlin and Vienna, and he would not be likely to have slanted his report in a way that would weaken his own case. Throughout his statement he sees Grey as envisaging a military conflict and wanting to localise it, though doubting his ability in certain circumstances to hold back Russia. Even on 20 July, though the situation was now changing, Grey still told Lichnowsky that he would try to localise the quarrel. 'The idea of a war between the Great Powers of Europe', he said, 'must be regarded as in all circumstances inadmissible'; and the German Foreign Office communicated this to the court of Vienna – communicated it as an assurance received from Britain. Again, on 20 July, Grey still showed himself set at very much the same angle in relation to Russia. He made a further attempt to reconcile the court of St Petersburg to an Austrian *démarche* against Serbia. The Serbian government may have been 'negligent', he says; and this time the excuse that seems to satisfy him is that the murder may have been 'planned on Servian territory'. Now, however, he adds a new point and says: 'It would be very desirable that Austria and Russia should discuss things together if they become difficult.'[42]

In these early weeks of July, the documents of the European powers – those of France, Austria, and Russia, for example – contain references to the way in which Great Britain has recently come closer to Germany, or has developed a friendship for Austria-Hungary.[43] Before the middle of July, the German ambassador, Lichnowsky, has begun arguing on his own account against the Austrian *démarche*;[44] but, in his own descriptions, Grey appears as more co-operative, more unprotesting, at this moment than he himself in the arguments he addressed to Berlin.

V

Sir Harold Nicolson, when he was discussing 'a punitive expedition', wrote: 'The Austrian government, most unfortunately, delayed taking action against Servia.' He added: 'It was this unexpected delay that upset the optimistic calculations of Downing Street and the Wilhelmstrasse.'[45] This view deserves to be seriously considered, for, after the early days of July, the effective new factor is the postponement of what was to have been the hot, indignant reaction of the court of Vienna to the assassination. You cannot claim the right of sudden retribution, or the prerogative of uncontrollable anger, if you have been coldly calculating your revenge and suspending passion during the course of nearly a month. One might almost say that the slowness of the Habsburg dynasty had caught up with them – Maria Theresa had long ago predicted that this defect in their system would prove their doom. While they were waiting in July, they had to bridle the press, calm public opinion, reassure the stock exchange, and carry out a prolonged process of deception. Many people became confident that this Habsburg dynasty would end by doing nothing; and after this kind of *détente* it was difficult for Europe to recapture the original indignation. On 9 July, Sir Arthur Nicolson (though, in this, he differed from Sir Edward Grey) had 'doubts as to whether Austria will take any action of a serious character'.[46]

But the original mood – the original expectation of a sudden, angry reaction on the part of Austria-Hungary – explains why Grey could be so unprotesting when he learned on 6 July that Austria might even attack Serbia and that Germany might not be prepared to hold her back. And some of the benefit of the same explanation should be given to the Kaiser not only when he gave the 'blank cheque' on 5 July but also when he committed what has been held to be a further offence. It is clear that if the Austro-Hungarian plan were adopted, speed – the swiftness of blind anger – was bound to be an essential feature of the case. The Kaiser knew his Habsburg allies and more than once urged them to act quickly. There was an occasion, nearly three weeks later (25 July) when Germany may seem to have behaved in a scandalous manner, still urging the Austrians to make haste. But, absurd though it might seem at that late date, the reason given is in order to prevent as far as possible the intervention of a third party'. The whole point of the action was to present the world with a *fait accompli*.[47] There is often an element of bluff in diplomacy; and the Kaiser on 5 July, though he thought that there would be no general war, certainly realised that he was risking such an eventuality. The Central Powers argued, however, that if, on the present occasion, the *Entente* declared against the Habsburgs, this meant that they were thoroughly hostile, thoroughly bent on war; and in such a case it would be better to have the conflict now. Furthermore, the Germans put on record their view that-granted this game of bluff – the best way to prevent the intervention of other great powers was to grasp the nettle hard, and show no sign of flinching.[48] They also insisted, however, that the moral sense of Europe was on the side of Austria and that the court of Vienna should keep it in that favourable state.[49]

By 24 July a number of things had changed, or the context around them had changed. And this partly explains why, from that date, there appears at the Foreign Office the idea that the Germans had been hoaxing the British. There is a trace of the same feeling in the Kaiser himself when, a few days later still, he returns from a three weeks' cruise. He, too, seems to realise that the Austrian government has saddled him with something different from what he had expected. The passage of time altered many things by changing their bearings. Whereas Germany had been working to secure the localisation of an Austro-Serbian conflict, there appears in our Foreign Office papers after 24 July the suggestion that she was simply a militaristic power, anxious to bring about a great war.

On 24 July two important events helped to produce a considerable change in the situation and policy of Great Britain. First of all, we learned the text of the ultimatum which Austria-Hungary had delivered to Serbia the day before;[50] and Grey described it as 'the most formidable document ever addressed by one State to another that was independent'. He now told the German ambassador that he would have no influence at St Peterburg if he tried any more to hold back Russia.[51] In fact, Eyre Crowe, when he first learned the character of the ultimatum, wrote a significant Minute in which he said: 'The localisation of the conflict has been made exceedingly difficult.'[52] We can take this as supporting evidence for the view that 'localisation' had hitherto been the aim of British policy. Grey knew the general character of the ultimatum when, the same day, and after a cabinet discussion of the Irish question, he asked his colleagues to stay behind for a few minutes and opened to them the problem of Austro-Serbian relations.

(Recent weeks had seen the climax of the Irish trouble, the threat of armed force in Ulster, and the unusual measure of George V, who summoned a conference of the leaders of all parties for a discussion of the whole question.) The Foreign Secretary, after calling attention to the European crisis, told the other ministers that 'he thought the position was very grave'. He evidently attached some importance to his suggestion that the courts of Vienna and St Petersburg should establish closer discussions, for he said that 'he was hopeful that the conversations which were proceeding between Austria and Russia might lead to a pacific settlement'. This was on a Friday evening. Lloyd George tells us that the Foreign Secretary left for his fishing lodge on the following day, and 'all the other ministers followed him and left town'.[53]

Secondly, up to this moment, Grey would seem to have been acting not so much as a representative of one of the opposing power *blocs*, but rather in his more recent role as the builder of a bridge between the two. He had handled the various states (Serbia excepted) as though they were all members of the same club. He was now called to order, however, by his other partners in the *Entente*. On the evening of 24 July, there arrived a long telegram from St Petersburg. It told how, during the recent visit of the French President, Poincaré, Russia and France had solidified their alliance with one another and renewed their mutual vows. Poincaré had in any case effectually thwarted Grey's desire that Russia and Austria-Hungary should hold direct discussions on the Serbian problem-a prospect which he described as 'dangerous'. It has been suggested that he feared the possible weakness of the Russians in conversations of this kind. Both Sazonow, the Russian Foreign Minister, and Poincaré, had already shown themselves bitterly hostile to any Austro-Hungarian idea of exploiting the Serajevo tragedy for political purposes. And one further result of the French President's stay in St Petersburg was bound to produce a peculiar headache for the British at this moment. Russia and France had 'decided to take action in Vienna with a view to the prevention of a demand for explanation or any summons equivalent to an intervention in the internal affairs of Servia'. This would have thwarted even a more moderate measure than the one actually taken by Austria-Hungary. The policy would have destroyed the whole hope of checking for the future the Pan-Serbian agitation and the conspiracies. Furthermore, Russia now called on Great Britain to 'proclaim [her] solidarity' with the other two Entente powers.[54]

The effect on the British Foreign Office was immediate. Eyre Crowe's Minute of 25 July on the report from Russia is one of the most significant that he ever penned.[55] It begins with a sentence which, once again, makes clear what the policy of the Foreign Office had been down to this time. Crowe says: 'The moment has passed when it might have been possible to enlist French support in an effort to hold back Russia.' He raises, then, an issue that is most important: he suggests that the rights and wrongs of the

Austro-Serbian conflict are now irrelevant. France and Russia have decided that the essential issue is the Triple *Entente versus* the Triple Alliance; and this means that the crisis is going to be taken as the test of Britain's fidelity to the *Entente*. If war actually breaks out, says Crowe, then, sooner or later, England is bound to be dragged into it; and we had better reconcile ourselves to this idea straight away. Grey himself shows that he is in a different mood now – he puts it on record that 'our attitude... will be regarded by Russia as a test'. Crowe added the point that, if Britain declared her solidarity with France and Russia, the Germans would give way; and in this he has been followed by a number of historians, who feel that Grey ought to have made the Germans realise Britain's determination to act with the other *Entente* powers. Grey did not accept this policy; but it should be noted that, by 31 July, Eyre Crowe identified himself with the ideas of the Foreign Secretary. He then declared that the British government 'should not by a declaration of unconditional solidarity with France and Russia... *induce and determine* these two Powers to choose the path of war'.[56] Grey was resolved to keep his freedom and independence as long as possible, but, now, *Realpolitik* made a stronger claim on him. The object of preventing a war had to be balanced against the need to see that if war came, it should be waged under advantageous conditions. Henceforward, so far as Grey was concerned, there was a sense in which the French and the Russians, in the extreme eventuality, would have the last word.

The beauty of Grey's diplomacy in the next fateful week springs from the fact that he was compelled to walk a dangerous tight-rope. If he allowed the Germans to count on Britain's determination to remain neutral, the Kaiser might be encouraged to refuse all concessions. But if he said too definitely that Britain would fight with France and Russia, he thought that they, too, might be tempted to behave provocatively. Even now, the virtues of his policy depended on his remaining free of continental commitments to this degree – his behaving still almost as though he were a free agent. But what forced him to walk this tight-rope, after Russia had demanded solidarity with the *Entente*, was his dependence on British opinion; and this gave him a handle against Russia herself. He knew that if he committed Great Britain unjustifiably, the nation or the Parliament or the cabinet would refuse to follow him. He was able to turn his political weakness at home into a diplomatic asset-able genuinely to say to both parties that the attitude of Great Britain would depend on the way they behaved.

Even when the crisis had become acute on 24 July, he did not instantly and entirely break away from the policy that he had been pursuing earlier. It was significant, for example, that he suggested at first a mediation between Russia and Austria – not a mediation between Austria and Serbia. He was still seeing himself as standing apart from Russia, and accepting the Austro-Serbian conflict as the given thing – trying to prevent the involvement of Russia. As the French historian, Renouvin, points out, he was still trying to localise the Habsburg war against the Serbs. As late as 29 July he made it clear that Britain would not want to be involved in the Austro-Serbian conflict as such or even the Austro-Russian conflict as such.[57]

We have seen how, nearly ten years earlier, an anti-German group had gained the ascendancy at the Foreign Office, chiefly through the activity of Sir Francis Bertie. Bertie himself, however, had apparently modified his attitude somewhat in the meantime and the first entry in his published Diary (an entry for 26 July 1914) presents us with quite a surprise. It is evident that, even at this late hour, he was still in the mood that Harold Nicolson described-still working on the policy and the assumptions of the immediate post-Serajevo period. Apparently unaware of the changes of 24 July, he writes two days later:

> It seems incredible that the Russian Government should plunge Europe into war in order to make themselves the protectors of the Servians.... And if she adheres to her present attitude she will expect France and England to support her by arms. Public opinion in England would never sanction such a policy; but unfortunately we might be dragged into a war through reverses to French arms and the necessity to prevent the annihilation of France.[58]

These ideas and assumptions appear again in telegrams, etc, from Bertie to the Foreign Office at this time.[59] Bertie's statement agrees in one point with Crowe's – he sees that in the event of war the involvement of Britain is likely to take place irrespective of the rights and wrongs of the Austro-Serbian quarrel.

VI

The lapse of time was itself to make an Austro-Hungarian attack on Serbia appear less and less like a mere 'punitive expedition'. It diminished also the chances of success – other powers were alerted in the antecedent stages, instead of finding themselves confronted with the *fait accompli*. In the passage of time between 8 and 24 July, Grey himself had been shifting his view somewhat – and showing himself less ready to think that Austria had a right to take revenge. There had always been another aspect of the Austrian design, however – the notion of what one might call, rather, preventive war. Apart from this, any attempt on the part of the Habsburgs to deal with Serbia could be construed as a political exploitation of the Serajevo tragedy.

The Kaiser seems to have held the view that military action in itself – though followed by immediate withdrawal – would alter the balance in the Balkans to the benefit of the Central Powers. But there remained the question of the possible annexation of Serbian territory; and though Austria denied any such intention – declaring her 'territorial disinterestedness' – there were bound to be doubts about this, and the Germans themselves had reason for wanting clarification. The Germans indeed became anxious about the matter since it became clear that there would be serious difficulties with Italy if annexation was envisaged. And, even if there were no such intention at first, nobody could be sure that the court of Vienna would not become more covetous once their troops were in Serbian territory. The Germans were anxious about Italy and presented the Austro-Hungarian government with reasons why an annexation policy would prove an embarrassment. It seems clear that they wanted to moderate the views of their ally, and if they did not ask or require moderation-if they merely inquired about Austria's intention-this was because they tried consistently to adhere to their policy of leaving the decision to the court of Vienna. That court appears to have intended only a slight rectification of frontier; but there seems to have been a serious deception involved in the assertion of 'territorial disinterestedness'. They had reverted to a scheme which they had meditated in earlier years – not taking Serbian territory for themselves, but dividing it between other states, enlarging Albania and Bulgaria in particular. The Austrian idea that the agitations and conspiracies could be brought to an end is hardly comprehensible in itself, but it acquires some meaning in the light of this contemplated partition policy. It was intended so to alter the map of the Balkans that a Greater Serbia agitation would be unrelated to anything in future. At this stage of the proceedings, the German Chancellor showed himself somewhat disturbed by the kind of duplicity in which the Austrian government was involving itself in the discussion of the question of territorial cessions.[60]

After 24 July, the European situation was different, the two *blocs* were now at grips with one another and, henceforward, Grey could only be 'semidetached'. He now had to consult France and Russia about some matters; and there were some things that he could not do in St Petersburg, some things that he could not do in Vienna, because Britain was a member of the *Entente*. His first intervention – his proposal of Austro-Russian conversations-had been vetoed by Poincaré as we have seen. His first idea of mediation was rejected by the other *Entente* powers because it was to be mediation between Austria and Russia – taking the Austro-Serbian conflict as a *fait accompli*, a thing to be localised. On 26 July, he accepted the *Entente* idea of a mediation having reference to the Austro-Serbian dispute itself; but now, on the 27th, it was the Germans who refused to co-operate, since they had promised Austria–Hungary that they would hold the ring for her and prevent the intervention of other powers. The position of the Central Powers was serious, for their prestige was at a low ebb after the assassination, and, so long as it had seemed that Austria-Hungary would do nothing, its weakness had come to be regarded with considerable contempt. A withdrawal by that power, after it had shown its hand and given the summons to Serbia, would have been more devastating in its results than any initial inactivity.[61] There is always a special problem in diplomacy when a state that is asked to accept a form of conference can calculate with something like certainty in advance that it will be outvoted if the conference actually meets. Germany and Austria-Hungary had not taken the Italians into their confidence, or approached them in the way Lichnowsky had approached the British Foreign Secretary. When the Italians got wind of Austria's intention they seem to have set out to provoke both Russia and Britain against it.[62] Their behaviour was far more hostile than Grey's had been. Italy would stand on the side of the *Entente* in any conference on the subject of Serbia.

Up to this time the Germans were still carrying on their game of bluff, trying to hold the ring for Austria-Hungary and hoping that if their ally would go firmly forward this would be the best way to escape the intervention of other powers. In fact they were merely adhering to the policy they had adopted, the resolution they had made on 5 July; and no influence from the military side is necessary for the explanation of this attitude, since, once the commitment had been made, withdrawal from it was a very serious matter. But by the evening of 27 July it became apparent that the bluff was not likely to be successful and that even Britain, whom they were closely watching at the moment, was taking a serious line. The German Chancellor telegraphed to Vienna at midnight: 'By refusing every proposition for mediation, we should be held responsible for the conflagration by the whole world, and be set forth as the original instigators of the war.'[63] But, even writing this, the German government were in terror of making it appear that they were deserting Austria, and they made no actual recommendation to her.

That night the Kaiser returned from his cruise and in the morning – 28 July – he read the Serbian reply to the Austrian ultimatum. He wrote in the margin of the document an entry which he can hardly have imagined would ever the published: 'A great moral victory for Vienna, but with it every reason for war drops away.... On the strength of this *I* should never have ordered mobilisation!' He at once wrote to the Secretary of State: 'I am convinced that on the whole the wishes of the Danubian Monarchy have been acceded to. The few reservations that Servia makes ... could be settled by negotiation.... *Every cause for war* falls to the ground.'[64]

Since the Serbs were so evasive, however, he still said that Austria must hold Belgrade as a hostage, as a guarantee that promises would be kept. The Chancellor proposed in fact to the court of Vienna 'a temporary occupation of Belgrade and certain other localities on Servian territory, in order to force the Servian Government to a complete fulfilment of its demands'. But always the Germans were afraid that Austria would think they were deserting her; and the ambassador who communicated the proposal had 'to avoid very carefully giving rise to the impression that we wish to hold Austria back'.[65] On the following day (the 29 July) when the attack on Serbia had begun, Grey himself put forward almost exactly the same proposal. He asked that there should be mediation on the basis that 'Austria, after occupying Belgrade, for example, or other places, should announce her intentions'.[66] It was almost a case of hauling the Austrians back to the notion of a demonstration, a punitive expedition, and inducing the powers to accept just this situation – to negotiate on the basis of that *fait accompli*. On 30 July the Chancellor wrote to Vienna: 'We are, of course, ready to fulfil the obligations of our alliance, but must decline to be drawn into a world conflagration by Vienna ... without having any regard paid to our counsel.' The Austrians took the line that if they merely made a demonstration and withdrew, Serbia would be left in much the same condition as before. Within two or three years the same trouble would be occurring again.

The process worked itself out unrelentingly: the Russian partial mobilization on 29 July, after Austria-Hungary had declared war on Serbia; the growing ascendancy of military considerations in diplomacy; then the solidarity of France and Russia in the face of Germany. What the Germans had hoped to secure from Grey was a localization of the contemplated Austro-Serbian conflict; and, for the accomplishment of this object, Grey would seem to have done the best that the situation would allow him to do. It could not be argued that any anti-German prejudice of his had limited his endeavour to prevent an Austro-Serbian conflict from leading to a general war. It is true that his policy was conditioned partly by the state of opinion in both the cabinet and the country. It is also true that his diplomacy – by the very fact that it left the invasion of Belgium to decide the issue – enabled Great Britain to enter the war with a very considerable degree of unanimity. The German government can hardly have been so naive as to rely to any great degree on Britain's remaining neutral once the invasion of Belgium had occurred. And it is not clear that Grey, who now wanted to intervene in the war, would have gained his object if the Belgians had decided not to resist invasion. Any serious criticism of British policy would have to go back to an earlier stage in the story – back to the time when it might be argued that Grey played his part in dividing Europe into two irreconcilable *blocs*.

VII

There is one area of the subject which has sometimes been neglected; and, where such a thing occurs, it is sometimes useful to direct one's attention to that place. Those who discuss the danger from Germany in 1914 tend to say little about Russia's projects of aggression; or leave them in the shadow, as something merely peripheral. It is generally true that, when the accepted story confronts one with an anomaly or a discrepant fact, it is a mistake to dismiss this as a mere irrelevancy or to attempt to poke it into the existing framework of story.

The anomaly is present, in 1914, if a man like Lord Morley can ask the question: And when you have dealt with Germany what are you going to do about Russia? We see it in those earlier critics of Sir Edward Grey, who resented not only the evils of Czarist Russia but also the definite and locatable aggressions of that power. It becomes particularly significant if we are interested in political morality; for, as Harold Temperley once said: 'I fail to see any respect in which Russia was better than Germany – and in some respects she was worse.'[67] The latest volume of *British Documents* to be published dealt with the period just before July 1914, bringing the story down to the edge of the ultimate crisis. It appeared only in 1938, and the coming of the Second World War, as well as the subsequent switch in our interests, may have prevented it from having its due influence among us. To one American reviewer the volume gave the impression that Britain by this time was more afraid of France and Russia than of Germany herself.[68] This reminds one of the place in the Memorandum of 1907 where Crowe confesses that there is one thing which would frighten him more than Germany, namely to have both France and Russia against us.[69]

The diplomatic documents do not suggest that either the British or the French or the Triple Alliance regarded the power of Russia at that time as illusory or anticipated the breakdown that it was soon to suffer. When Sir Edward Grey was in Paris in 1914, he found people tremendously impressed by the colossal power and the potentialities of Russia. He told the French that, in the event of war, he thought that the British would regard Russia, with all her resources, as capable of reducing Germany to exhaustion without their help.[70] During the crisis of July 1914, the British Foreign Office received a dispatch from Paris containing significant news about Russia's resources, and showing how, by 1916, her army would exceed that of Germany and her allies combined. Knowing what actually happened to Russia during the First World War, I imagined, when I first read the document, that the leaders of the British Foreign Office would be sceptical about this kind of assessment and would show their scepticism in scathing Minutes. The only Minute that has been reproduced by Gooch and Temperley, however, is one by Sir Arthur Nicolson who wrote: 'Russia is a formidable Power and will become increasingly strong.' He then added: 'Let us hope our relations... will continue to be friendly' – not the usual comment of British diplomacy when a colossus is arising.[71] One wonders what was expected to happen if a victory over Germany at that time brought Russia to the top of the world. In recent months, that power had driven the British to exasperation in Persia – this at a time when, far from being at the top of the world, she was actually courting us, trying to commit us to a definite alliance, and telling herself that she would be invulnerable if only she could be sure of the British Navy. She did make one interesting offer: the idea of a mutual guarantee of Asiatic possessions – an agreement which might be extended, she said, to Japan.[72] This offer gave the British some hope at first; for it might mean at least a paper-guarantee of India. But then there was a doubt about it; for it could hardly provide satisfaction in respect of Persia. I do not know what significance to attach to the fact that this proposal for a *détente* in Asia was put forward on 9 July 1914.

* * *

By 1914 it was apparent that there were two giants in Europe: Germany and Russia. When there are two of these monsters, however, the threat to liberty is not doubled; it is likely to be greatly reduced. Each ought in a sense to be precious to all the rest, who, if they cannot prevent a war between them, ought to dread most of all the kind of war that will eliminate one of them, while bringing the other to the top of the world. In those days, nobody could know that, after a war in which Germany was broken, Russia herself would be eliminated – unable to share in the fruits of victory or to secure that the promise of Constantinople should be made good. It is possible, therefore, that only the emergence of a new factor, the Russian Revolution – a thing not embraced in the calculations of 1914 – helped to conceal the fallacies of pre-1914 diplomacy.

Fear of the Russian 'bear' had been a recurring feature in history, but, save among the Germans, it had often been fear of a contingent danger – fear of the colossus that Russia was likely to become. To some people, Czarist Russia represented the darker threat, and was the more inexorable of the two dangers. Once she achieved the ascendancy, she would have the better chance of rendering it permanent, the better chance of securing that a broken rival should never be able to reconstitute its power. But, to many Westerners, the danger from Germany was at any rate the imminent one, and domination from that quarter would be the more immediately unpleasant. If Britain thought that Russia would soon be as formidable as Germany, still there might be a serious danger arising out of the time-lag – a danger that the Germans would force the issue before the Russians were ready. A flexible diplomacy is able to address itself to a dual problem of this kind: for it is not a rare thing to have a menace urgently impending from one side, while a remoter threat, not less terrible, looms up from the opposite quarter. The trouble is that, even though sometimes it would relieve them of many headaches, people do not like to have to think of the existence of two dangers (or two potential enemies) at the same time. Indeed, the fear which Germany provoked in the twentieth century made many people (but not all) forget their older terror of the Russian 'bear'.

Professor R. W. Seton-Watson, in his presidential address to the Royal Historical Society in 1946, made the remark: 'The Second World War, still more than that of 1914–18, was essentially a Slav war – the greatest Slav war in all history.'[73]

This carries a suggestion which is unfair to the intentions of Grey in 1914; for Grey, who in any case had little patience with Serbia, never faced the question of Russian domination *versus* German domination. He stated on

29 July 1914 that Great Britain was not disposed to interfere in the conflict between Austria–Hungary and Serbia. He added (very understandably) that Great Britain would not want to interfere even in a struggle between Teuton and Slav for supremacy in the Balkans.[74] Indeed, it is not clear that he worked from any comprehensive map of European forces, or with any historical sense for the relationship between Germany and Russia. What interested him was the western European problem, and, most of all, the question of the relations between Germany and France. Like Crowe and Bertie he held that Britain could not on any account afford to see France overrun by the Germans; and this point seems to have been regarded as axiomatic. Strictly speaking, such a maxim of policy was calculated to override the whole question of the rights and wrongs of whatever controversy had brought France and Germany into a state of war with one another in the first place. Grey, Crowe, and Bertie certainly believed that it would have to be brought into operation even in the case of a war in which Britain had originally refused to take part. Perhaps the maxim itself, which seems to have been of pivotal importance, has never been sufficiently examined. One would like to know in any case whether certain irrational factors were not involved in the British attachment to France. At any rate, the doctrine may have helped to restrict Grey's effective horizon and prevent him from embracing in his vision the whole field of European forces.

In the years before 1914, it is not clear that, between the rival giants and the rival continental groups, Grey sufficiently realised the elements of strength in Britain's diplomatic position. Powers that recognised their great need for his support were too easily able to make him feel that England was in danger of being without a friend in the world. Between 1912 and 1914, Britain's need of the continental states was nothing like the need of both the opposing groups for something or other that she had to give or to sell. And if England could not afford to see French territory overrun by Germans, it was necessary for her to keep a certain control over the development of the situation, and to avoid being frightened by the argument that she might be considered not worth having as an ally. In the years before 1914, only a little would have been needed to prevent the dreadful hardening of the situation – perhaps not more than the difference between what actually happened and what some of Grey's opponents in the cabinet thought the situation to be. Crowe's famous Memorandum is remarkable for the way it treats Germany in isolation, without dealing with Russia or surveying the European scene in any comprehensive manner; and this was perhaps understandable at a date so soon after Russia's defeat by Japan, and the subsequent revolutionary outbreaks. In any case, an assessment of the situation made in 1907 could not have the same validity in 1912; and here was a further case for flexibility – for a policy that would prevent the alliance-groups from becoming rigid, irrespective of changes in the balance of forces. England, if she had been only a little less clearly committed, might have provided for some years the one loop-hole for the admission of greater flexibility into the diplomatic situation. Grey almost managed to open the loop-hole early in July 1914, and that crisis of July is sufficient in itself to show that a bridge between the two groups could be a matter of importance. Grey's work for peace in the final crisis tended in reality to be constricted by his obligations to the *Entente*, and it was rather helped than hindered by the fact that he could point to the pressure of public opinion, or the resistance of his colleagues, who did not recognise the commitments in the way that he did. His persuasions might not have been different but his actual *will* might have had greater weight with both the Triple Alliance and the *Entente* powers in 1914, but for certain aspects of his earlier policy for which he rather than the cabinet had been responsible.

It could not even be said that, before 1914, while the German threat was the imminent one, the threat from Russia was merely contingent or hypothetical. The ambitions in the Balkans, the determination to have Constantinople, and the advances in Persia were already a factor in the story. But Grey found a way of sliding the Russian aggressions into a pocket, so that they did not complicate his main picture. The fact that it was not regarded as feasible for Britain to meet the threat to her Asiatic interests by a great display of force was one of the reasons for the conclusion of a convention with Russia. It soon had to be recognised, however, that even the *Entente* did not put an end to the activities in Asia to which Britain objected. In the last resort, it was Grey's claim that Russia's offences had at least been moderated and reduced because of the existence of the *Entente*. In other words, the aggressions would have been more serious, and the conflict more acute, if England had not been Russia's friend. This may be collated with the Foreign Office Minute which stated that Britain must expect to suffer in India if the Russians and their allies defeated Germany without her assistance. The Minute carries with it, however, the suggestion that if Britain had been the ally of Russia during the war itself, the gratitude of that power would make it more accommodating when it found itself at the top of the world.[75] Britain, in all this, was adopting a special attitude to the aggressions of Russia because she was 'on our side'. This enabled her to write off a considerable area of the diplomatic problem, and avoid comparing power with power, aggression with aggression – avoid properly confronting the map of European forces as a whole. It left Germany standing as simply the supremely wicked ogre, the condemnation of which was not to be offset by any proofs that Russia had dangerous ambitions and meant to achieve them by military force. For the same reason, British diplomats (and later historians) were unwilling or unable to see that Germany might have a genuine fear of Russia, and that this might account for some elements of intransigence in her policy that seemed unreasonable. British unimaginativeness about the German fear, and indeed about the danger of a Russian ascendancy, is one of the curious features of the story – a thing calculated at times to add a certain desperateness to German policy. Though the fact is difficult to reconcile with his general attitude, Grey, at the beginning of July 1914, did in reality provide signal evidence of his belief in the reality of the German fear.[76] It is hardly the Germans who have the right to blame him, for (influenced by a militaristic interpretation of their own history which did less than justice to the element of reasonableness in Frederick the Great and Bismarck) they failed to pursue the kind of diplomacy that the occasion had come to require. They are much to blame, therefore, if it was out of their power to make that margin of difference which might well have saved both themselves and the balance of Europe when the menace of Russia later confronted the world unmistakably. Grey's whole policy would be more explicable (but his intellectual system more difficult to square with itself) if it could be said that his real and ultimate fear was the fear of what would happen to India if Russia were not 'on our side'. Some of the Foreign Office Minutes that have been mentioned half reveal and half conceal this lurking fear; but, as Nicolson put it, one hoped that this formidable power would be 'friendly', one must be 'most careful not to alienate her'.[77] Diplomacy is no longer diplomacy if, when aggression is in question, the issue pivots finally on such an act of faith or such a piece of sentimentality.

Notes

* Originally published in: *Historical Studies*, 5 (1965), 1–25. Reprinted by permission (eds).

1. *The Times*, 22 August 1914.
2. See Max Beloff, *Lucien Wolf and the Anglo-Russian Entente, 1907–1915* (London, 1951), 35–6. cf. G. M. Trevelyan, *Grey of Falloden*, 254 n.
3. Temperley MSS.
4. G. P. Gooch, *Under Six Reigns* (London, 1958), 88; 64.
5. *Cambridge Historical Journal*, ix. 25l: 'Notes on the Origins of the War of 1914'. by Harold Temperley, dated 30 December 1927: 'It is, I think, a grave step to assume that the conclusions of either German writers or militarists (General Staff etc.) are identical with those of any German Government... I don't think it can be truly said that the militarists got control until about the 30th July 1914.'
6. G. P. Gooch, op. cit.,175.
7. Viscount Grey of Falloden, *Twenty-Five Years, 1892–7916* (London, 1926), ch. 2.
8. [Gooch and Temperley] *Brit[ish] Doc[ument]s [On the Origins of the War, 1898–1914]*, iii. 397–420. (See paper 4 in this volume). (eds).
9. Spenser Wilkinson, *Thirty-Five Years, 1874–1909* (London, 1933), pp. 187, 221, and 316–19, which compares parallel passages in Crowe's *Memorandum*, and his own *National Awakening*.
10. *Brit. Docs.*, iii. 420–31. Cf. Z. Steiner, 'The Last Years of the Old Foreign Office 1898-1905', *Historical Journal*, vi. 79.
11. Dissertation by G. W. Monger, 'The End of Isolation: British Foreign Policy, 1901–1907', which is in course of publication. Cf. *Historical Journal*, vi (1963), 78–83, and ibid., p. 59, which shows that the change was all the more important inthat 'after 1906, men who had been clerks began acting as true advisers'. Cf. p. 88. The first division clerks 'became advisers engaged in the policy-making process.... The

new senior officials were men with definite political aims', and 'Grey was young, relatively inexperienced'.

12. Eugen Weber, *The Nationalist Revival in France, 1905–1914* (University of California Press, 1959).
13. *Brit. Docs.*, viii. 399–411, e.g. p. 401, dispatch of the British Military Attaché in Brussels, 8 October 1912, where the Belgian War Minister is reported as saying: 'Would Great Britain come to the assistance of Belgium if her neutrality were violated by the French? He thought not'; and pp. 405 and 408, Annual Report for 1912, where the same attaché writes:

    The Government attitude appears to be that England must be looked upon as an ally of France... and an interested power against whom defence measures must be taken and plans of campaign prepared'; and 'There is now an impression which the Government have perhaps created and are assuredly cultivating that the position as to Belgian neutrality has changed since the establishment of the *entente* with France.

14. (A. Doren) *Amtliche Aktenstücke (zur Geschichte der Europaïschen Politik 1885–1914)* (Berlin, 1925), v. 249–51, Extracts from a Report from St Petersburg, 27 March 1914; 259–61, Report from Paris, 8 May 1914; 273–7, Report from Berlin, 12 June 1914; 277–9, Report from Paris, 24 June 1914.
15. *Brit. Docs.*, x, part 1.360: Sir L. Malletto to Sir E. Grey, 5 December 1913. Cf. ibid., pp. 367–8, same to same, 10 December 1913:

    The Admiral has the absolute command of the Turkish fleet in time of peace.... The concession just obtained, giving us control over the dockyards and arsenals for thirty years, makes it all the more difficult for us to take a strong line. In practice, moreover, I believe that the command of the 1st Army Corps [by Liman von Sanders] will be entirely innocuous to our interests.

16. Ibid., pp. 360, 362, same to same, 5 and 8 December 1913. See also in n. 6 of the latter page the minute of R. G. Vansittart: 'The effect of now curtailing Admiral Limpus' powers w[oul]d certainly have the effect foreshadowed by Sir L. Mallet.'
17. Ibid., pp. 365, 371, 383.
18. Ibid., pp. 365, 371. Cf. ibid., pp. 381–2, H. J. O'Beirne to Grey, 14 December 1913:

    'I am afraid that, unless we satisfy the Russian Government... it is certain that they will revise their estimate of the value to them of their present understanding with us.... If His Majesty's Government feels unable to follow Russia... we must, I believe, be prepared for a definite change in the attitude of the Russian Government towards us'; ibid., p. 409, Sir G. Buchanan to Grey, 6 January 1914:

    His Excellency... intimated that if England failed to support Russia after all pacific means had been tried and failed there would be an end to our understanding'; and ibid., 414, same to same 7 January 1914: 'It will constitute the touchstone by which the value of the French alliance and of English friendship will be tested by Russian public opinion.

19. Ibid., p. 389, O'Beirne to Grey, 18 December 1913.
20. Ibid., p. 407. Cf. ibid., p. 423 Grey to Buchanan, 11 February 1914: 'I think that the intrinsic importance of [the question] has been very much exaggerated.'
21. *Preussische Jahrbücher*, clvi (June 1914). 394. The Russian historian in question was Professor Mitrofanof.

22. M. N. Pokrovski, *Pages d'Histoire* (Paris, 1929), p. 128–32, showing the uneasiness of Sazonow at the reports from Giers in Constantinople to the effect that the Turks wanted an alliance with Russia; with the instruction to Giers, so August 1914: 'Ayez en vue que nous ne redoutons pas une entrée en action de la Turquie contre nous.' The Germans were equally uneasy at the possibility of having Turkey as an ally, requiring defence against Russia: Kautsky, etc, [Die] Deutschen Dokumente [zum Kriegsausbruch] (Charlottenburg, 1919), i, 70, 98, 136 (correspondence between Berlin and Constantinople, 14-22 July 1914). Cf. L. Bittner and Hans Uebersberger, *Ö[sterreich]-U[ngarns] A[ussenpolitik]* (Vienna and Leipzig, 1930), viii. 441, 451, 461, 477, 530–2. Cf. ibid., p. 705.
23. Ibid., 821, Sir A. Nicolson to Buchanan, 14 July 1914.
24. [O.] Hoetzsch [*Die Internationalen Beziehungen im Zeitalter des Imperialismus*], series I (Berlin, 1932), iv. 2–7, Benckendorff to Sazonow, 28 June 1914.
25. *Brit. Docs.*, iii. 418.
26. Lord Hardinge, *Old Diplomacy* (London, 1947), p. 159, 176.
27. *Brit. Docs.*, xi. 5; Grey to Sir E. Goschen, 24 June 1914. Cf. *D[eutschen] D[okumente]*, i. 8, Lichnowsky to Bethmann-Hollweg, 24 June 1914. There is a translation of this latter work, *Outbreak of the World War, German Documents collected for Karl Kautsky*, produced by the Carnegie Endowment for International Peace (New York, 1924).
28. *Cambridge Historical Journal*, ix. 254–5.
29. *John Bull*, xi July 1914. Cf. Ö.-U.A., viii. 364, Bericht aus London, 8 July: 'Die sache im höchsten Grade apokgryph schien.'
30. *Brit. Docs.*, xi. 130, Grey to Sir M. de Bunsen, 27 July.
31. H. Nicolson, *Lord Carnock* (London, 1930), p. 410.
32. *D[ocuments] D[iplomatiques] F[rançais]*, 3rd series, x. 793 n.
33. Winston Churchill, *The World Crisis, 1911–1914* (London, 1923), p. 193. Cf. ibid., p. 199: 'On Monday [27 July] began the first of the Cabinets on the European situation which thereafter continued daily, twice a day.' Lloyd George, *War Memoirs* (London, 1936), i. 53.
34. *Brit. Docs.*, xi. 25–6, to de Bunsen.
35. See n. 23 above.
36. Loc. cit.
37. *Brit. Docs.*, xi. 24–5, Grey to Sir H. Rumbold, 6 July. Cf. *D.D.*, i. 40–2, Lichnowsky to Bethmann-Hollweg, 6 July. Lichnowsky says that he told Grey that

> the Imperial Government could not be blamed for refusing to let this new challenge pass or for demanding satisfaction from the Serbian government, in view of the demonstrable support which the conspirators had received from Belgrade. Whether, and in what form this would be required, I did not know, but I believed that it would be advisable just now to have in mind the possibility of a sharpening of the relations between Vienna and Belgrade, so that he, Sir Edward, would be in a position at the proper time to exercise his influence at St. Petersburg, with the object of persuading Serbia to submit to the Austrian demands. Sir Edward…did not fail to recognise the danger that the situation might produce and appeared to understand that it would be difficult for a leading Austro-Hungarian statesman to refrain for any length of time from energetic measures. He promised that he would keep in touch with us on this matter, but for the time being declined to express any definite opinion.…He insisted…that it was his object to bring the two groups closer together.

38. *Brit. Docs.*, xi. 30, Grey to Bertie, 8 July.
39. Ibid., xi. 30–3 i, Grey to Buchanan, 8 July; Hoetzsch, i, part iv. 141–4, Benckendorff to Sazonow, g July.
40. *Brit. Docs.*, xi. 33–4, Grey to Rumbold, 9 July. Cf. *D.D.*, I, 51–2, Lichnowsky to Bethmann-Hollweg, 9 July, according to which '[Grey said that] England wished to keep an absolutely free hand so that in the case of European complications she could act in accordance with her own estimate of the situation. The Government had to a certain degree committed itself to Parliament not to make secret engagements that would tie its hands.... All the same, relations with [France and Russia] had lost none of their earlier intimacy.... He seems ready to try in one way or another to influence the attitude of Russia if we wish it. In case the Vienna Cabinet feels compelled to adopt a sterner attitude to Serbia as a result of the murder at Serajevo, he has already been trying to persuade the Russian government to take the matter quietly and assume a more conciliatory attitude towards Austria. Very much would certainly depend, in his view, on the kind of measures that are under consideration and on the question whether they arouse Slav sentiment in such a manner as to make it impossible for Sazonow to remain passive. In general the Minister was in a thoroughly confident mood and declared in a cheerful tone that he saw no reason for being pessimistic about the situation.
41. *D.D.*, i. 77–8, Lichnowsky to Jagow, 20 July. Grey's account is in *Brit. Docs.*, xi. 54, to Rumbold, 20 July. 'I said ... I hated the idea of a war between any of the Great Powers, and that any of them should be dragged into a war by Servia would be detestable.'
42. Ibid., p. 53–4, to Buchanan, 20 July.
43. *D.D.F.*, 3rd series, X. 730, Dumaine to Viviani, 13 July; Jules Cambon to Bienvenu-Martin, 20 July. *Ö.–U.A.*, viii. 408, Bericht aus Berlin, 12 July. Cf., however, ibid., 280, Bericht aus London, 3 July, where Menadorif writes: 'I hear from the Germans that they are nervous in Russia about London and Berlin coming together', but explains that, unless Russia makes great mistakes Grey will not change his policy of friendship with that country. Cf. also *D.D.*, i. 16–17, Tschirschky to Bethmann-Hollweg, 2 July.
44. *D. D.*, i. 68–9, 78, 82, 88–90, dispatches and letters from Lichnowsky, 14–16 July. The last of these contains a most serious and interesting criticism of the general policy of Austria-Hungary. There is a reply by private letter from Jagow, 18 July, ibid., pp. 99–101.
45. H. Nicolson, *Lord Cammock*, 410.
46. *Brit. Docs.*, xi. 33, Minute to dispatch from Bunsen of 5 July. Cf. ibid., xi. 59, Minute of Nicolson to dispatch from Bunsen of 21 July. 'I doubt if Austria will proceed to any extreme measures.' In *D.D.F.*, 3rd series, v. 792, Paul Cambon (to Bienvenu-Martin, 22 July) describes Nicolson as 'plus optimiste que Sir E. Grey'. According to Cambon, Nicolson said that all the representatives of foreign powers in Vienna, and especially the French Ambassador, believed that Austria would do nothing. When Grey on 8 July warned the French Ambassador of possible Austrian action, the report of the latter (ibid., 691–2) shows that he did not regard the possibility as a serious one. In *D.D.*, i. 98, Wangenheim reports to Berlin from Constantinople, 18 July, that people in that city, especially in *entente* circles, were refusing to take seriously the idea of an Austrian *démarche*. In Hoetzsch, I. iv. 297, Benckendorff (to Sazonow, 22 July) writes that Grey is very disturbed, but 'Nicolson sagte mir, er teile diese Beunruhigung nicht in demselben Grade.'

47. *D.D.*, i. 216, Tschirschky to Jagow, 26 July. Cf. *O.- U.A.*, viii. 704, Szögyény to Berchtold, 25 July.
48. *D.D.*, i. 100, Jagow to Lichnowsky, 18 July: 'The more determined Austria shows herself, the more vigorously we support her, the more chance there will be of Russia remaining quiet'; ibid., p. 129, Von Jagow to Chargé d'Affaires in Hamburg, 22 July: 'There is no better way of avoiding war than by our taking our stand quietly but firmly alongside Austria–Hungary as a matter of course.'
49. Ibid., ii. 39–40, Bethmann-Hollweg to Tschirschky, 28 July:

> 'In the event of a completely uncompromising attitude on the part of the Austro–Hungarian government, it will be necessary to reckon with a gradual defection of public opinion throughout Europe from its cause.... If it continues to maintain its previous aloofness... it will incur the odium of responsibility for a world war.... A successful war on three fronts cannot be undertaken and maintained on such a footing. It is a commanding necessity that the responsibility for a possible extension of the war... should under all circumstances fall on Russia.... The public opinion of all Europe... is on the point of turning away from Austria.'

Also: *Ö.-U.A.*, viii. 408, Szögyény to Berchtold, 12 July.
50. *Brit. Docs.*, xi. 73, Grey to Bunsen, 24 July. Mensdorff had told Grey the nature of the ultimatum the day before, ibid., p. 70, same to same, 23 July.
51. Ibid., p. 73, Grey to Bunsen, 24 July; and p. 78, same to Rumbold, 24 July.
52. Ibid., p. 79, Minute of Sir Eyre Crowe, 25 July, to communication from the German Ambassador of 24 July urgently calling for a localization of the conflict.
53. Lloyd George, op. cit., i. 54.
54. *Brit. Docs.*, xi. 80, Buchanan to Grey, 24 July.
55. Ibid., pp. 81–2.
56. Ibid., p. 201, Minute to dispatch from Bertie of 30 July. Crowe still held, however, that 'If and when... it is certain that France and Russia cannot avoid the war, and are going into it,... British interests require us to take our place beside them as Allies.'
57. Ibid., p. 180, to Bertie.
58. *Diary of Lord Bertie* (London, 1924), i. I.
59. *Brit. Docs.*, xi. 97. Bertie to Grey, as July. 'I expressed... this afternoon the opinion that public opinion in England would not sanction a war in support of Russia if she, as protector of Slavs, picked a quarrel... over Austro-Servian difficulty.' Ibid., 99, same to same, 25 July. 'I do not think that if Russia picks a quarrel with Austria over Austro-Servian difficulty, public opinion in France would be in favour of backing up Russia in so bad a cause.' Ibid., 133, same to same. Private. 27 July: 'I am sure that the French Government do not want to fight and they should be encouraged to put pressure on the Russian Government not to assume the absurd and obsolete attitude of Russia being the protectress of all Slav states whatever their conduct.'
60. *D.D.*, ii. 58, e.g. Bethmann-Hollweg to Jagow, 29 July.
61. See *D.D.F.*, 3rd series, v. 789. 'Notes sur les rapports austro-serbes', dated Vienna, 17 July, and communicated by the General Staff of the army, where it is submitted that, if there is no *démarche*,'this is only conceivable if Austria-Hungary abdicates and admits the annulment of her prestige or if she has been warned by her allies that she will not be supported in any way. Nous ne la croyons pas déchue à ce point.'

62. *Ö.-U.A.*, viii. 494, Tel. aus Rom, 18 July, with footnote; ibid., 521, Tagesbericht über einen Besuch des deutschen Botschafters, 20 July; Hoetzsch, I, iv. 225–6, Tagesaufzeichnung des russischen Aussenministeriums, 16 July.
63. *D.D.*, i. 268.
64. Ibid., i. 264; ii, 18.
65. Ibid., ii 39–40, to Tschirschky, 28 July. Cf. ibid., p. 125, same to same 30 July.
66. *Brit. Docs.*, xi. 182, to Goschen. The file copy of this is marked: 'Not sent – War.' Cf. *D.D.*, ii. 86, Lichnowsky to Jagow, 29 July, and 124, Bethmann-Hollweg to Tschirschky, 30 July. Also *Brit. Docs.*, xi. 196–7, to Buchanan, 30 July, a telegram communicated to Lichnowsky, and through him to the German Foreign Office, *D.D.*, ii. 181.
67. *Cambridge Historical Journal*, ix. 252.
68. *Journal of Modern History*, X (1938), R. J. Sontag, 'British Policy in 1913–14'.
69. *Brit. Docs.*, iii. 417.
70. Ibid., x, part ii, 788, Grey to Sir F. Bertie, 1 May 1914. Cf. the Belgian reports, n. 14 above.
71. Ibid., Xl. 53, minuting dispatch from Lord Granville, 18 July. Cf. *D.D.*, ii. 72–3, where Lichnowsky reports that Sir W. Tyrrell (after Grey, 'the most influential as well as the most instructed person' at the Foreign Office), says that 'nach ihren Nachrichten' the Triple Alliance 'would not be able to stand the test of a world war'. Cf. also *Journals and Letters of Reginald Viscount Esher*, iii (London, 1938), 175, a letter from Esher under the date 5 August 1914: 'If the German Emperor is Napoleon the campaign of 1814 will repeat itself. If he is not, he is done a month hence when Russia advances.'
72. Ibid., xi, xi, Buchanan's private letter to Nicolson, 9 July; same to Grey, received 19 July; Grey to Buchanan, 20 July. Cf. ibid., X, part ii, 820 f. Cf. no. 24 above.
73. *Transactions of the Royal Historical Society*, 4th series, xxix (1947). 185.
74. Ibid., xi. 180, Grey to Bertie, 29 July. However, he said, 'If Germany became involved and France became involved, we had not made up our minds what we should do.'
75. See n. 55, above.
76. See pp. 9–10, above, and n. 39.
77. See n. 71, above.

Part II

Faith and Ethics in International Politics

7
Introduction

A lifelong Christian, Butterfield (together with R. G. Collingwood and M. Oakeshott) had an important influence on the 'Cambridge School' of political theorists who looked to history, philosophy and law rather than logical positivism – the precepts of Social Science – as the key to sound political enquiry. A profound understanding of earlier generations, they believed, was essential if one was to manage with a reasonable degree of success, the complexities of the present. Underlying Butterfield's conception of international politics (and history) was his persistent distinction between secular history and Divine providence, between the 'City of God' and the 'city of man' – a distinction whose implications was an abiding ordering principle for his views on the human past and his construction of historical narratives. The Methodist foundation of Butterfield's faith with its Wesleyan admonition 'there but for the Grace of God go I', made him perennially sensitive to the dangers of self-righteousness, intolerance and zealotry in political life; averse, in other words, to secular absolutes: extremist doctrines and their attendant systems.[1]

Augustinian in his convictions, Butterfield never believed in any 'permanent political solutions' to human problems; he never saw the secular realm, its tragedies, absurdities and aspirations in anything but transitory contingent terms. At best historical events, ideally tempered by religious convictions, comprised the reservoir from which we can distil the insights that will better enable us to understand public affairs, both in their domestic and international dimension.[2] Statesmanship at any level required the faith, skill and judgment derived from accumulated experience, guided by Divine inspiration – what Butterfield called 'true wisdom', in order to be effective. Among his most significant services to historical scholarship was the ongoing quest for a rediscovery and enhanced understanding of God: revealing both the eternal truths of religion and its enduring significance for society and the world (religion here being distinguished from theology and its doctrinal forms; for Butterfield it was intended to elucidate the timeless aspirations of mankind and revivify a sense of the sacred and transcendent).

Butterfield also warned that history could easily become the tool of politics – a weapon in ideological conflicts and danger to a free and civilised society. He persistently argued against personifying ideas, against seeing them as self-explanatory, autonomous agencies in history – just as he resisted the projection of the State as creator and arbiter of human values, priorities and historical truths. By depicting the dangers inherent in the deification of the State, he saw Christianity as a defence against totalitarianism, the worship of secularised collectivities and moralistic judgments in politics, which, historically, have fostered endless hatreds, suffering and destruction. The earnest quest for utopias seemed, to him, an absurdity, as did the investing of ideologies and political causes with visionary commitments. Only the workings of Providence, guiding temporal forces, injected truly meaningful pattern and design into the grand sweep of universal history.[3] A concomitant challenge, at all levels, according to Butterfield, was the problem of 'politicisation' – the prevailing characteristic of our age. This entails making all questions political questions; all issues political issues; all values political values; all decisions political decisions – factors governing every relationship between people in society and resulting in correspondingly mounting pressures within the international sphere once both individuals and state entities become engulfed by the regimentation of centralised power.[4] These developments, in turn, have been intensified in the 20th century by renewed visions of the millennium; of that better world which intellectuals promote and which politicians promise but never deliver. Unfulfilled hopes, Butterfield argues, and ideologically stimulated passion in the face of seemingly anti-progressive forces, are the effective ingredients of violence and endemic malaise of all modern societies: not only did he see the world becoming more 'political' but also as becoming more passionately political with an intensity rarely, if ever, previously experienced in history. Butterfield countered this tendency with an uncompromising intellectual integrity, humility, candour and deep spiritual commitment – though not without an unblinkered ability to look facts squarely in the face: what has caused scholars to associate him with the prominent tradition known as 'Christian realism'.[5] Ever pragmatic despite his persuasive moral approach to domestic and international life, Butterfield clearly perceived the incongruity between political norms and those underlying Christian ethics. In politics, norms necessarily 'play a more circumscribed role than in the lives of individuals: the ethics of limitation supplanted the ethics of aspiration'.[6] As such, his ideas were less prescriptive than admonitory: advocating a more flexible attitude towards history – the processes of time – together with a more imaginative understanding of the complex dilemmas in political affairs and the means with which to face them.

The following papers illuminate various aspects of the themes discussed above and have a shared concern – one that permeates Butterfield's international thought as a whole: namely, to demonstrate that Christianity, in its spiritual and historical dimensions, offers authoritative, practical precepts for meeting contemporary needs worldwide; the wisdom of former ages *can* be transposed into modern terms. Although Christianity and the Western moral tradition, as is argued in the first essay, cannot emancipate human existence from its tragic boundaries, an eternal predicament, they provide more opportunities for creative solutions than raw power politics based on national egotism, aggression and expediency. Christianity, according to Butterfield, does this, in part, not only because it obliges believers to transcend conventional assumptions – habitual, worldly ways of thinking – but also because it views all conflicts in terms of 'universal cupidity', human nature being defective since the Fall of Man. For Butterfield, international crises were not an apocalyptic collision between forces of 'good and evil' but a mundane, tangled admixture of sins and misperceptions on all sides, with often disproportionately catastrophic consequences: 'deadlock between one half-right that is too willful and another half-right that is too proud'.[7] In this scenario, even Christians are merely confessed sinners; not the vindicated 'righteous' arraigned against the 'wicked', a fundamental precept that resonates throughout the first paper, showing that precisely because of their transcendent loyalties, Christians have it in their power 'to be more flexible in respect of all subordinate matters' and to ally themselves 'with whatever may be the best for the world at a given moment'. Therefore, a Christian 'is particularly called to carry his thinking outside that framework which a nation or a political party or a social system or an accepted régime or a mundane ideology provides'. Such independence of thought, seasoned by Christian teaching – the notions of humility and divine love above all – prompts the believer to be skeptical about the presumption 'that his own nation's cause is absolutely the righteous one and all wickedness on the side of the enemy'. This fosters a timely awareness of contingency, frailty and incertitude within the human condition – a perception which, in turn, serves as a safeguard against partisanship and intransigence; promoting that capacity for strategic perceptiveness and moderation so conducive to wise statesmanship. Concurrently, he reiterated a point already disturbingly made in *Christianity and History*: the small transgressions of ordinary individuals, when multiplied a million fold, can produce greater calamities than all the wars waged by the Hitlers of this world.

The second paper 'Human Nature and Human Culpability' emphasises the theme of Christian love, or in his words, 'Christian charity' as an access to those qualities of mercy, tolerance, and compassion which are indispensable conditions of a workable international order, however transitory given the human condition.[8] 'These qualities would help us, Christian or non Christian, from being hideously moral in a way that is calculated to be ruinous to the world' – calculated to make human relations an endless cycle of uncomprehending hatred, vindictiveness and retaliation. Butterfield was disquietingly realistic in his appraisal of human nature – not to mention the nature of states – throughout time: what he advocated was not idealism or altruistic moral regeneration but practical, rational safeguards against human excesses; the disposition of institutions and structures designed to curb delinquencies and temptations. Nurtured by Christian values, these can also soften what Butterfield calls 'modern barbarism' – a phenomenon becoming more widespread even in developed nations, 'where the legacy of Christianity and the traditions of humanism have been violently dispersed', with serious global schisms and breakdowns the usual result.[9] Barbarism, however defined, needs to be met with more than its own methods, a reminder of timely relevance for our own present 'war on terror'.

Although Butterfield was writing against the background of American–Soviet nuclear tensions, his prescriptions carry a timeless quality: future issues and combatants may be different yet the dynamics of wars will remain the same if they are driven by ideologies grounded in dogma and moralism. The greatest menace to civilisation past, present and future, he cautions, is 'the conflict between giant organised systems of self-righteousness – each system only too delighted to find that the other is wicked – each only too glad that the sins give it the pretext for still deeper hatred and animosity'.

Here Butterfield, as explored in the third paper below – relied not only on the simplicities of faith but also, more pragmatically, on a 'scientific' rather than 'moralistic' approach to human affairs – 'scientific' signifying an experiential field brought under intellectual/rational control – or made amenable to it – as opposed to one wholly engulfed within the realm of passion, bigotry and prejudice. Ultimately the 'scientific' treatment of international power and war analyses 'predicaments' rather than persons and ideologies, structures rather than the moral responsibility of political leaders or the meaning of the ideologies articulated by either side.[10] Instead of conflicting systems, which are mutually exclusive, it was more constructive, argued Butterfield, to construct an international order based on positive interactions and the reciprocity of interests. This is the purpose which transcends national egoisms and puts the boundary to self-interest – the purpose to which all our more immediate aims in foreign policy have reference. He expands upon this in paper four, 'Moral Judgments in History' where he urges policy-makers to jettison their private views and personal valuation in order to concentrate on the 'tangible evidence' – the 'kind of truths which the evidence forces us to believe whether we like them or not'. This prevents moral judgments from becoming political ones in disguise or from working themselves into the very fabric of historical thinking and reconstruction. Parties, regimes and ideologies have an innate tendency to partial explanation – one that ratifies and exonerates their own versions of events – but halts short of genuine understanding, letting moral condemnation, ethical censure and resultant national animosities do the rest. The result again, according to Butterfield, is the kind of impasse invariably created by the very structure of human interactions without extenuating causes. In an imperfect world it is *love* and *charity* that must be the operative criteria. This complements the social order described in his famous *Christianity and History*, now transposed into the international realm: an order comprised of subtle inducements and stratagems designed to harmonise the clash of political wills and shifting configurations of power.[11] Butterfield clearly realised, as illustrated in paper five, 'The Moral Framework of International Relations', that the moral propensity of states is bound up with the operation of force – a connection that has remained constant throughout history regardless of the nations involved. Wilful expansion, even by previously non-aggressive states, is an ever threatening danger unless countervailing structures and principles are in place. Here he extolled, at times almost nostalgically, the 18th century experience which, with its conscious attempts to curb the evils of war through creative diplomatic initiatives and conventions, managed to fashion a *raison de système* conducive to compromise and a viable international order.[12]–an order serving moral purposes by placing civilised restraints on the potential anarchy of nation states. One critical component of this, in Butterfield's opinion, was the deepening understanding of individual personality and human relations generally – a notion central to his overall historical methodology, ethical orientation and view of Christian responsibilities in the political/secular world.

Notes

1. On this see most recently, McIntire, *passim*: M. Thorp, *Herbert Butterfield and the Reinterpretation of the Christian Historical Perspective* (Lewiston, NY, 1977), chs I, II. M. Hobart, 'History and Religion in the Thought of Herbert Butterfield', *Journal of the History of Ideas*, vol. 32 (1971), pp. 54–54. C. T. McIntire, ed. *Herbert Butterfield: Writings on Christianity and History* (New York, 1979). R. Epp, *The Augustinian Movement in International Politics: Niebuhr, Butterfield, Wight and the Reclaiming of a Tradition*. Occasional Research Papers (Aberystwyth, 1991).
2. Coll. op.cit., 4–13, 33–57. K. W. Thompson. 'Butterfield and the Theory of International Politics' in: Thompson ed. *Herbert Butterfield: The Ethics of History and Politics* (Lanham, MD) 41–54. H. Butterfield, 'International Relations', Butterfield Papers (CUL) 1980, File 295.
3. See *inter alia*: H. Butterfield, *Christianity and History* (New York, 1950); idem, *Christianity in European History* (London, 1951); idem, *Christianity, Diplomacy and War* (London, 1953); idem, *International Conflict in the Twentieth Century: A Christian View* (New York, 1960); idem, 'Conflict and Ethics', BC File 285; idem, 'The Nature of Political History', ibid., File 132.
4. K. S. Templeton ed., *The Politicization of Society* (Indianapolis, 1979) pp. 169–207, 211–247.
5. Coll, op.cit., p. 3, E. Patterson ed., *The Christian Realists* (Lanham, MD, 2003), pp. 1–25, 8–102; J. Derry, 'Herbert Butterfield and the Interpretation of History', English *Historical Review*, CXXI (2006), pp. 25–60.
6. Coll, op.cit., p. 5.
7. H. Butterfield, 'Tendencies in Historical Studies in England', *Irish Historical Studies*, 4:15 (1945), 219–23; idem, 'Morality and Process in International Affairs', BC File 31, cf. Thorp, op.cit., pp. 99–119.

8. For an elaboration of this important point see: Malcolm Thorp 'The Inescapable Predicament: Sir Herbert Butterfield's Reflections on the Human Dilemma', *Fides et Historia*, vol. 16 (autumn 1983), pp. 7–17.
9. M. Bentley, 'Herbert Butterfield and the Ethics of History', *History and Theory*, 44 (Feb. 2005) pp. 55–71.
10. McIntire, op.cit., 244. Herbert Butterfield, Letter to Peter Savigear, 16 Feb. 1970. BC File 18.
11. As elaborated in: M. Keens-Soper, 'The Practice of a States System', in: K. W. Schweizer ed., *Diplomatic Thought 1648–1815* (Sherbrooke, 1982) ch. I.
12. H. Butterfield, 'The New Diplomacy and Historical Diplomacy', BC File 329; idem, 'The Development of Diplomacy', Ibid., File 137.

8
Christianity and Human Problems[1]

There are a number of reasons why those who try to do their thinking as Christians may legitimately hope to make some contribution toward the solving of the world's problems. And though the attempt may be unsuccessful on any particular occasion, the same reasons always impose upon the Christian the duty of trying to break through the conventional framework of contemporary thinking on those subjects.

In the first place there have accumulated around the Christian tradition various forms of worldly wisdom which condense the experience of centuries and have come to stand as part of our European heritage. For nearly two thousand years our religion has particularly represented and upheld the element of continuity in our civilisation. Even in matters that are of merely mundane concern, the historical Church has tended to cherish the long-term precepts, the truths that span the ages. It is true that the weight of so long a past – the drag of so much history which Christianity carries behind it – has had many disadvantages, especially at the very point where religion plays its part in public affairs. A tradition misunderstood, or inflexibly interpreted, is no doubt always one of the most sinister of the dangers which can menace an historical Church. Yet in the business of a nation, or a European order, or a civilised world, it is fatal to disregard one particular kind of tradition – namely, those maxims of conduct which embody the results of man's long-term experience on the earth. A certain sagacity in this part of our Christian heritage – though occasionally it has been regarded by churchmen almost as evidence for the rightness of the religion itself – has owed some of its virtue to the workings of time and the continuity of a single experience which ripens and corrects itself through a succession of centuries. The benefits of this accumulated wisdom can never be gathered by the mere reading of history books; for what a man gets out of history depends only too often upon what he first puts into it; and one can so re-write the story of mankind that all its tendency is to ratify those errors which happen to be current at the present day. On the other hand, it is true that many of the maxims which were once associated with the Christian outlook underwent secularisation at the time when the world began to shake itself free of religion. Those which concern international affairs came down to us therefore rather as traditions of European diplomacy, and we have ended by losing sight of some of the Christian assumptions which really lay behind them. Let us say, then, that the 20th century has made some serious breaches – and among these possibly some desirable breaches – with the traditions of European diplomacy, for example. The Christian – still somewhat the guardian of the continuities in our civilisation – may be concerned to inquire whether anything valuable is being thrown away at the same time.

At the opposite extreme we may ask in the second place whether, in spite of appearances, Christianity is not eminently capable of producing the greatest breach of tradition – precipitating the unpredictable thing. We may wonder whether the Christian does not find himself in the middle of a set of relationships calculated to catapult him to some interesting point in space where new perspectives will give him a glimpse of new things. If he is attached to long-term traditions the world is apt to lock itself into short-term ones, and is slow to break through the intellectual framework of a mere generation. Its thoughts often go on revolving only within that framework, and turning in on themselves, while men vainly imagine that they are producing something new. The world, moreover, trembles as though confronted with the spectre of bankruptcy if the intellectual framework even shows signs of breaking down as a result of its own defects. Precisely because he can hold fast to spiritual truths – not turning any mundane programme or temporal ideal into the absolute of absolutes – the Christian has it in his power to be more flexible in respect of all subordinate matters, and to ally himself with whatever may be the best for the world at a given moment. In the last resort he can clear away all intermediate systems and meet an unprecedented problem by going back to his starting-point – his zeal for personalities as such. It is open to him to unload himself of all the intervening years and even to cut through the traditions of historical Christianity, so that all may be fluid and flexible save that ultimate Rock, which is Christ Himself. He may, for example, return to the Sermon on the Mount or the Gospel narrative, and confront the twentieth century with the original simplicities of the faith. In the time of the greatest stress and the darkest tragedy this may be the only means by which human values themselves are likely to be re-established. Even among men who would hardly claim to be numbered with original thinkers, the policy of referring back to the primary Christian truths may precipitate new things and prove to be a starting-point of historical change. In the 16th century those who insisted that Christ had come to make men free helped to secure that modern sectarianism should be the parent of political liberty. In a similar context the ideals of communism and egalitarianism were introduced to the modern world.

Like our predecessors in so many generations, we are often tempted to think that the order of things in which we live – or some such revision of it as we have already compassed in our imagination, so that we can almost reach it with our fingertips – will always be the best thing possible for the world, preserving just those values to which we are attached at the present moment. Like our forefathers in the age of aristocracies, we cannot believe that an order of things which we have not imagined or promoted may bring new values, new patterns of human life, and new benefits to the human race, under a set of conditions that is beyond our foresight. Yet while men weep to see the end of the things they love, something in Christianity survives the fall of Roman Empires and national monarchies, lives on when Platonism and Aristotelianism go out of fashion, and persists when a whole civilisation changes its character. The Christian is particularly called to carry his thinking outside that framework which a nation or a political party or a social system or an accepted regime or a mundane ideology provides. Even the preservation of what we may love as the existing order of things – even Anglo-Saxon ideals and western values – are not the absolute absolutes for the Christian. It is always possible, therefore, that a return to the first principles of our religion will once again precipitate upon the world a new thing.

It may be argued, however, that if traditionalism can run to excess, originality itself and the breach with tradition – the leap into the unpredictable – possess perils of their own. And no doubt Christianity in its essence is a risky religion, packed with the kind of ethical implications that are dangerous to *status quos*, established regimes, and reigning systems. Perhaps we ought to impute a far more considerable role in history than we usually do – and ascribe a far more damaging influence – to those people who perform the function of making Christianity a safe religion, accommodating it to the existing order of the world, and rendering it more compliant toward the powers that be. They are the agency through which the Church came to be hated as the instrument of tyranny; and they would seem to have operated as one of the primary causes of its decline. Even within the normal routine of historical Christianity, however, there is the basis for a third reason why the Christian should attempt to be independent in thought, and should let his Christianity do his thinking for him. Certain aspects of Christian teaching – the treatment of love, the insistence on humility, the attitude to human personality and the doctrine of sin – have helped to affect the character of our civilization, no doubt , but still leave a divergence between the message of the Church and the assumptions which are current in the world. Through direct reference to such a teaching, many Christians, for example, have played a distinguished part by their care for the suffering – even the suffering of their enemies – in time of war.

In general, I think that in time of war we should expect the Christian to have compassion somewhere even for the enemy, and even for the wicked – expect him also to be diffident about believing that his own nation's cause is absolutely the righteous one, and all the wickedness on the side of the enemy. And, important though it always is to win a war, we should expect the Christian to be diffident about believing that the fate of God and all His angels in all the future is going to depend on the Greeks beating the Trojans in one age of history, or on Russia rather than Germany dominating the continent in another age of history. The Christian has principles which can rescue him from the blindness of mere partisanship – principles which come like a compassionate wing to cover the whole of this wilful and distracted human race. Without being either traditionalist or revolutionary, but merely by following a course that is currently recognised – following a principle to which all men do lip-service – he has a chance of performing a role of the highest strategic importance at the present day.

In *The Listener* for 21 September 1950, Professor James Welch examines the position of those Africans 'who still need the best the British have to offer, and equally resent needing it'. He particularises the case which the Africans make against the British, and enumerates the complaints which we have to raise in return; and the result is a startling example of argument at a deadlock, argument at cross-purposes. He continues:

> Only a few days ago I heard a West African putting before a sympathetic group of British the feelings and convictions of his people and the reasons for their hostility to Britain; and I heard him answered by a typical and sincere Englishman who is entirely devoted to the African and wholly given to serving him. And as I heard that good Englishman, from the best of motives, criticise Africans for what he believed was their own good, I was filled with despair. For... all his desire to help the African was of no avail in making him feel with the African in what he had been saying. The gulf between the two remained. Neither was able to move a step toward the other. Communication had broken down.

And the author of this article asks the question which is bound to preoccupy all Christians at the present moment:

> Is there a creative way out of this *impasse*? Can a bridge be built between the two points of view sincerely held by sincere people?... Is there a way of life, a philosophy, which is essentially personalist, giving primacy of importance to human persons, and, at the same time, free from racial and colour conflict and prejudice?... No problem of human relations, however intractable it may seem, is insoluble to genuine Christian love.

When Mr Welch speaks of the *impasse* to which we come in our relations with the West Africans, we can add, within the lifetime of most of us, the parallel cases of the Irish, the Egyptians, the Indians, the Malayans, and others. In international relations the deadlocks are only too liable to be of precisely the same kind; and before any powers take the decision to resort to war there is a point where the argument has come to a similar *impasse*, and there is a gulf which human imagination has failed to bridge.

In view of his awareness of the streak of egotism in human beings, no Christian will dare to assert that all the world's quarrels are to be solved by human understanding. Human understanding is urgently needed as a preliminary, however, and it is essential to recognise that fact that there are problems of human relations which the intellect of man has never yet been able to grasp. Throughout history we see dark screens that stand as barriers to international comprehension – we see sincere men, only too conscious of their good intentions, but locked in their separate worlds and arguing woefully at cross-purposes. If all men were to become as unselfish as St Francis, and were to do so in time to remove the present causes of stumbling, the problems would be solved, and a profound Christian message would be vindicated. Short of that, however, it is our duty to see that life in the world shall be as tolerable as we can make it; and some of the problems that are involved in the relations between peoples are problems addressed to our thought and our imagination. Here is an area of human life not yet brought under intellectual control, but amenable to it sooner or later if we try to exceed our present state. It seems to be the case that even the historian, remote from our passions and preoccupations, may need a long period before he masters the particular intellectual hurdles which our generation failed to surmount. I personally have no doubt that he will ultimately get over them, and perhaps even – though often too glibly – point out what wiser courses we might have pursued. I imagine that the thing the general historian would say about the barriers to international understanding would be that they go much more deep than is generally assumed, so that at any given moment we underestimate the effort that is needed to secure their dislodgement. We underestimate how far we have come short of what is necessary for a genuine understanding of the other party's case. There is urgent need for us to convince ourselves, however, that mere historical knowledge does not ever suffice to surmount the obstacle. As many of us know, a long period of service, however self-sacrificing, in India or in Africa may still leave a man in an attitude of 'Blimpishness'. It is not sufficient to have researchers in Germany or gatherers of information in Russia. We may listen as the African himself makes his explanations, and we still may not understand, because we are unable to feel with him – we do not give something of ourselves in order to achieve real comprehension. And in fact we all know that the barriers can exist between parties within the same nation, and even a man and wife may be blocked in the effort of understanding. There is a sense in which it would be proper, therefore, to describe most of our thinking in these fields as still somewhat at the pre-Copernican stage; for we have not sufficiently examined, or sufficiently allowed for, that totally unconscious bias in the vision of both the Englishman and the foreigner – the bias which arises from what I can only call the position and the state of the observer.

The problem is one, therefore, which calls not merely for knowledge but also for imagination. And like the proper study of history it requires a certain giving of ourselves – requires in fact that we shall do something with our personalities. What society needs is every possible variation and extension of the art of putting oneself – and actually feeling oneself – in the other person's place. And this process, which is the most difficult in the world, has an unfortunate way of appearing to be the easiest, especially to the honest man who is a partisan. Those who are farthest from possessing the capacity to achieve it are – by that self-same defect – the ones most incapable of realising the fact. Yet a person who is inadequate in this – however brilliant he may be in other ways – is totally incompetent for the solution of our particular problem. While we perfect our atomic bombs we are not always aware that we are required with equal urgency to increase our very stature – for it cannot be too often stated that it is human understanding itself which needs to be enlarged. And since it means thinking with our sympathies, and actually feeling with the other party – means giving something of our personalities so that we may comprehend the men not like-minded with ourselves – it is to Christian thought in particular that the problem is most likely to be presented. Little by little we may hope that Christian thought will turn into communicable knowledge those forms of intellectual exploration which are accessible only to men in a certain frame of mind, to human beings in love, human beings willing to make fools of themselves for love.

A comparatively small number of communists, possessing initially no measurable degree of power, achieved in the thirty years after 1917 such a transformation of the globe as has rarely been equalled in magnitude or in thoroughness or in extent. They achieved far more than all the armies of Kaiser and Tsar, of British and Austro-Hungarian Empires, and of the United States, put together. Above all, what they achieved was to a far greater degree what they had intended than had been the case with any of the alliances or the governments which were in operation at that time. Indeed, with the partial exception of that other rebel, De Valera, no other statesman or party then functioning ever came so near to producing the kind of world offered in a political programme or promised to the public – came so near to keeping control over the consequences of policy.

All those people who are driven to pessimism and paralysis by the idea that nothing can be affected on this earth save by numbers and power, may take comfort, therefore, from this exhilarating example. The reading of abridged text-books has the grave defect of preventing men from realising the extraordinary flexibility of the workings of history and the subtlety of the processes of time. Given the required intensity and intentness, a comparatively few Christians could alter the course of history as powerfully as the communists have done; and in fact this has already happened. It happened through the acquisition of power at a certain point in the story, however; and the power takes a terrible revenge on both the communism and the Christianity that lay hold on it. It might be said that the best Christian influence has always worked in history in a less spectacular manner, and has therefore had less tendency to provoke those counter-balancing evils which are responsible for the seamy side of the story. The profoundest effects on our civilisation have been produced by people who were not thinking of affecting the course of human affairs at the mundane level; they have come as the unanticipated by-products of lives that had really been intent on spiritual things. The highest achievement is not that which takes the world by storm after the manner of modern communism; but that which, quietly penetrating free minds, acts as the leaven which leavens the whole lump.

It might be thought, therefore, that the Christian should withdraw from the tumult of international affairs, the whole sorry story of power, diplomacy, armaments, and war; and should even require of his country the same policy of non-co-operation. But those who disarm when a colossus like Russia maintains its forces are handing the world over to power, and may be permitting the extension of the area of conquest, violence, and suffering; for there is in Christianity no promise that a miracle will prevent a Russia from making use of her advantage in such an eventuality as this. If an individual consents to make a self-sacrifice – even to face martyrdom before a foreign invader – it is not clear that he has a socially recognisable right to offer the same sacrifice on behalf of all his fellow-citizens, or to impose such self-abnegation on the rest of his society. Certainly we can rejoice in one truth: there is something so impressive in the power which gentleness can have in the long run, that passive suffering and the willingness to be a martyr seem ultimately to move the world more than the resolution which meets force with force. And in the worst crisis, if renewed war breeds more wars, and civilisation disappears in a thousand local conflicts, a state of misery could conceivably be reached in which the protest and the voluntary suffering of the pacifist might be the only lantern for the re-discovery of even the things which we call human values; provided the pacifism were consistent in its simplicity of mind – not the sort which can accompany over-vigorous action in the economic struggle for existence – not the sort which tries to dominate society by the formation of pressure groups. Wherever this form of witness authentically emerges, therefore, I presume that Christians would protect it in vindication of conscience, and guard it as the king of treasure which keeps its value when all prudential calculations fail.

But at the present moment vast accumulations of military force – colossal slabs of power – are actually in existence all around us. Let us not close our eyes to this fact, any more than students close their eyes to the workings of cupidity in economic life. We may say that the power would dissolve, as mountains of ice dissolve away in the sun, if all men were to become Christian saints; but Christians, besides working for remoter purposes, have to look for those intermediate measures of relief which the condition of the world requires in the meantime. We cannot say that power would melt away if the world became formally and officially Christian after the manner of former ages; for, fifteen hundred years of Christian predominance in Europe failed to stop wars, failed to eliminate this problem of power. All we can ask – while the military force heaps itself up around us – is the question: Can the world be made more tolerable in spite of this power which solidifies in great masses among nations and empires? Since the power does not disappear merely because we wish it away or because we cry out that it ought not to be there, is there anything in our way of thinking about it which can mitigate its effectiveness or limit its role in history? We need not attach ourselves to anything which has been thought or attempted in the past; but if our predecessors found ways of moderating power or cushioning its effects, we must not feel that all these are to be swept away by a brush of the hand – there is a wisdom of former ages the cream of which, at least, is often capable of being transposed into modern terms. And if predecessors of ours had particular success in the establishment of a comparative stability in their world, and at the same time saw that a genuine international order required the maintenance of certain delicate conditions, their statements are not to be dismissed with a mere catchword about the dead past –especially by a generation which has so palpably failed in this respect. Above all, if the conclusions at which we arrive are more conformable to Christianity itself – more consistent with charity, more exacting in self-discipline, and more conducive to a Christian spirit even in time of war – than the ideas of the contemporary world; and if, to boot, they offer a foothold for a more scientific thinking, a more systematic reflection on such of our conditioning circumstances as might be amenable to intellectual control – then, to judge by analogous cases, we have found a point where Christianity can get a purchase on human history, and can insert the influences which have worked to the same effect in other fields. Only after the last armed State has dismissed the last army – if even then – will it be anachronistic to discuss this ever-present issue, or prudent to allow the matter to fall into neglect.

Note

1. Orginally published in *Christianity, Diplomacy and War* (New York, 1953) pp. 1–13.

9
Human Nature and Human Culpability[1]

Even if our New Testament were a work of fiction, and were presented to us as the achievement of an imaginative poet, its disturbing challenges and its exposure of the human plight would still be as valid as ever they were, and as vivid as ever a Shakespeare could make them. And those who do not believe that a Jesus ever rose from the dead must recognise that somebody has left a pattern there, which has the effect of convicting every one of us.

It is staggering, however, to see the elementary character of the ideas which appear to prevail in the world on the subject of morality in general, and particularly on the question of human guilt and human culpability. Sometimes it would be less harmful if it was really believed that men are like automatic machines, the new products of heredity and conditioning circumstance, so that sin would be only a disease or maladjustment. The non-Christians do not adopt this attitude, however, at any rate when they are considering other people's sins. On the contrary, irreligious though they may be, they tend to be cruelly, grimly, appallingly, and even hideously moral, in a way that is calculated to be ruinous to the world. They tend to be moral in a manner that is actually stultifying; and indeed they are often not moral at all, but only moralistic. Just the thing that the pagans do not do is to make allowances for other people's sins, allowances for heredity and environment, for education and predicament, for favourable opportunities on the one hand and the constricting force of circumstances on the other. They do not even make allowances for the sins which they themselves would commit if they were in the other person's place. Partly because it asserts the sinfulness of all men, Christianity, in its doctrine of Love, goes beyond any other system in its recognition of all those points in which it is the duty of human beings to make allowances for one another. It goes further than any other system in its vision of what there is which is worth saving in every human creature; and though it brings a consciousness of sin, it refuses to see any man as necessarily the slave of his own past. It does not encourage but destroys what one might describe as that paralysing feeling of guilt which men sometimes desire to impose upon a defeated enemy.

It is a paradox that the highest and most spiritual view of life which is available to man – and the one which caries human beings to the most elevated and rarefied realms of experience – is one which starts with the assertion of universal human sinfulness. The very ladder which has carried men to those exalted spheres and regions of light has an end which rests in the mundane realm on the primary recognition of this fact. The finest examples of human sainthood and the finest blossoming of human personality seem to emerge out of an initial abasement of the human being before this very truth. They seem indeed to be inseparable from a continuous confession of sin, and the very power which works with such efficacy in holy people is the knowledge of the forgiveness of sins. It is hardly necessary to say that the recognition of all men as sinners is calculated to have momentous effects upon the whole world of human relations. It would be the Christian assertion, I feel sure, that human relations can never be properly envisaged – nor, of course, could any science of them be possible – except after this situation has been squarely faced. Even if we pictured Christians as set against the rest of the world, we should say that the Christians were the confessed sinners, not that they were the righteous arrayed against the wicked. In any case, we should not take the line that ours was a warfare against the rest, but rather that it was warfare on their behalf. From the time of the Gospels, indeed, it is precisely the self-righteous who are the enemies of the spirit of Christ. And it is a pity that the modern world has lost so much of the moral teaching which is to be drawn from Christ's controversies with the Pharisees.

But the greatest menace to our civilisation today is the conflict between giant organised systems of self-righteousness – each system only too delighted to find that the other is wicked – each only too glad that the sins give it the pretext for still deeper hatred and animosity. The effect of the whole situation is barbarising, since both sides take the wickedness of the other as the pretext for insults, atrocities, and loathing; and each side feels that its own severities are not vicious at all, but simply punitive acts and laudable measures of judgment. In this connection, let us note that if the world ever drops the notion of sin as a crime against God and replaces it by the notion of sin as a crime against man, we shall be setting civilisation back by far over two thousand years, back to a date far earlier than the life of Christ. For the present-day world, as it makes this transition, is not only turning its back on Christ, but is turning its back on humanising ideas about the nature of sin which can be traced to great religious movements in Asia hundreds of years earlier still. If man arrogates to himself the right which he is not fitted to possess, and which Christianity withholds from him – namely, the right to judge actual wickedness, to adjudicate on sinners and to punish the sin itself – then there can be no end to the atrocities. For man without God is terrible to those whom he chooses to regard as sinners; and he has become more terrible since he lost the idea of the Judgment of God, so that he dare not leave anything to the Judgment Day, for fear that this will never take place, and the sin will go for ever unpunished.

All this does not mean that Christians are to tolerate cruelty and suffering; for our endeavour in this respect has to stop short only at the limit of what is practicable or possible. It means rather that our endeavour to improve the world is not to be distracted or interrupted by considerations of justice that are purely of the vindictive kind. At this point, as at many others, the world is liable to overlook the strategic significance of the Christian principle of Love. When Francis Bacon urged the case for a new kind of knowledge which should be more capable of being harnessed to human utility, he used that Christian doctrine itself against what he regarded as mere intellectual indulgence on the part of the Aristotelians. On this basis he insisted upon the importance of a form of thinking which would relieve the lot of human beings. If, when all things are considered, we can diminish suffering, or can better train a child, by leaving an offence unpunished, then there is no blind judgement of heaven that will be warped to all eternity by our charitableness of mind. If we can win a man from a delinquency by treating him as no worse than other people, there is not an angry God who will say that He has been cheated in the winning of this soul. And the Christian, conscious of so many sins unpunished, is not the person who will insist on punishment when punishment is actually harmful. Nor will he fill the world with the uncomprehending resentments of men who – as in 1914 – knew little of the crimes imputed to them, and indeed suffered chiefly perhaps because they were the heirs of the system of Bismarck. Above all, if we really wish to enter into the fundamentals of the case, we should note that the real challenge to human sin is the challenge which itself springs from love. The case for the moralities can never be met if the culprits are being pilloried merely by political enemies, and if communists are charged by Englishmen in a manner calculated to improve nobody's lot in Russia, and only to generate more malice here. There is a 'war for righteousness' in the universe, but it is not fought at the level at which it serves the militant needs of one state against another. Even ecclesiastical systems which have made manifest the virtues of charity, are sometimes shown up badly in history when, instead of winning men, they proudly condemn – now trouncing the vices of democratic leaders without mentioning the virtues; and now showing more fervour in condemning sins abroad than sins nearer at hand. Even the Church is only invincible when it denounces man's universal sin – that is to say when it brings my own sins home to me.

And, curiously enough, it is not clear that on a last analysis, the historian, commenting on the evils of our time, will judge that the real source of them was any unusual criminal – even a criminal like Hitler himself. Criminal natures are likely to be found in all countries, and a healthy, stable order of things is likely to keep them in their place. If a man like Hitler gained a position from which he had the power to do magnified harm, it is quite conceivable that he never could have reached it but for the minor sins and small delinquencies and petty lapses from duty on the part of multitudes of men – multitudes not merely in Germany, but in other countries. And these multitudes were men, who, indeed, may have had their willful moments, but who never felt themselves to have been doing any great harm. And, similarly, if in one age aggression is more terrible and violent than in another age that is not to be explained on the argument that human nature is being produced on the earth in baser forms now than ever before. It reflects a failure in respect of that whole system of safeguards which has to exist in times of stability – safeguards against the cupidity of states, preventing men from feeling that circumstances give them the opening for a high gamble or a colossal design. And when all the responsibilities are levelled out and properly distributed, even the shallowness of modern idealists and the egotisms of the system-makers are not likely to be exempt. Furthermore, all this explains why the historian, as time recedes, sees the emergence of a Napoleon or a Hitler as an historical process which can be scientifically analysed.

There are vast numbers of people who can be regarded as neither absolutely vicious nor utterly unselfish, neither criminals nor saints. And some show a fine face to the world while they live in a conventional society, and have a smooth going-on, with only small temptations, and no storms but storms in a tea cup. Virtue in such cases is remarkably dependent on combinations of circumstances; and this must be one aspect of the Fall – namely, the fact that human conduct is so subject to conditions. Many will be reasonably virtuous if the conditions are not too hostile and the price is not too high; and things are so disposed that all the arrangements of a stable, healthy society tend to keep these people on the rails. Once there is a police strike, or a revolution, or a terrible dilemma, we are astonished to see how ugly this ordinary human nature can be; and it is then that we learn how greatly respectability had been buttressed by institutions.

These are the people – neither black nor white – who, when their nations are at war, will be passionately self-righteous, burning with indignation against one another. If the issue were plainly put before them they would shudder and shrink from the idea of dropping an atom-bomb at Hiroshima for the purpose of turning a negotiated peace into the luxury of 'total surrender'. But when such issues are presented to them – as they always are – somewhat muddied and complicated, they can see reason and right (observing the act from the point of view of the doer) in the very thing which would outrage them if it were done by the enemy and seen only from the outside. Alternatively, they can be sincerely horrified by an atrocity imputed to General Franco. But when the same atrocity is attributed to the Greek communists or to an ally of their own, they can be equally sincere in blotting out the culpability – blotting out everything from their minds save the allowances to be made, or the inner reasons which render the action more understandable.

Now it is the experience of history and the teaching of historical Christianity that something can be done even with this world of half-lights, this grey, drab universe of the half-righteous. And it is precisely because their conduct responds so greatly to conditions that we are able to have a science of society or a scientific history, and can show the processes of a revolution, or examine the development of a class-war, or estimate the probable reaction of public opinion to a given policy. It is against these people – against men like ourselves – that we raise up our mighty indignations when our country is at war. Yet the Christian attitude to them is not merely the higher charity; it is the only attitude that is consonant with a scientific treatment of human problems. Indeed, the great enemy in these human concerns – and particularly in the problems that relate to war – is the tendency to myth and eschatology. And the greatest advance was made in the treatment of the issue of war itself, in an age which rejected myth and eschatology at least in this particular branch of human affairs – an age which saw human nature as one, and was aware of its defects, but which, in these very defects, saw the grounds for a scientific attitude. For there could be no science of human affairs if all human beings were like gods, reigning unconditioned; or if there were no laws formulating the general effects of even ordinary egotism and wilfulness.

One of the greatest problems of the present day is the curious phenomenon which we have described as 'modern barbarism'. It is the kind of thing which, for understandable reasons, existed in the Balkans before 1914, and was long discernible even in the politics of that peninsula. It is the question of a kind of barbarism which can exist amid all the external apparatus of a civilised world; so that we may find it in countries which have suffered a reversion through wars, breakdowns, revolutions, disorders, or conditions of unusual pressure and desperation; but also we may find it in countries which have assimilated the science and technology, but not the subtler virtues of civilisation – countries which have copied democratic institutions without realising the tolerances and urbanities and self-discipline which they require. At the root of the whole phenomenon is a hiatus in the transmission of the higher values of a civilisation, a defect in the way in which the traditions of a society have been handed down. Often it is the imponderable virtues of a culture which are allowed to evaporate unless men take care of them; so that what we are really dealing with is the case of 'the disinherited', the case of people who have been deprived of an essential portion of their legacy. We even know that partial manifestations of the phenomenon may appear in our own country, it, owing to a war, a younger generation is allowed to miss an essential part of the teaching that is due to it. The same effect might be produced if, owing to a materialism in our outlook, we were to allow a purely technical training to supercede the education in human relations and human values. Our own world still possesses much humanity in spite of its departure from faith; for it is full of men who are virtually lapsed Christians, soaked in the traditions of our religion and the ideals of the Renaissance. But even in our own world we can begin to see what happens when a legacy so distinct and clear as this begins to melt away in the ocean of materialism. We can measure in some of the countries of Europe, therefore, the relations between cause and effect in the phenomenon of modern barbarism, where the legacy of Christianity and the traditions of humanism have been violently dispersed.

Wherever modern barbarism emerges we find ourselves shocked to see one general consequence. It is the disappearance of the tolerances and urbanities and self-discipline which we associated with a civilised world. Occasionally we find in history that the transmission of these things to a new class of society – previously 'disinherited' also to a certain degree – may become of momentous importance at a time when there is a redistribution of power inside a given country. Behind the evils that we deplore lies the failure to respect the other man's personality, the absence of the process of looking inside the other man's mind – there is an impulsive way of deciding that the man who disagrees with one's views is a rogue or a fool or has a vested interest. And in proportion as men have missed the training and the legacy which belong to the higher aspect of civilisation, this failure in the respect for human personality seems to become the more complete in its barbarity, and it is the quality of mercy which withers from the world. And *ipso facto* all conflicts – even party conflicts within the State and even personal quarrels – take on the blind uncomprehending character of 'wars for righteousness'. Whether modern barbarism in its most brutalised and uncivilised form occurs in Germany or in Russia, in Palestine or in China, in revolutionised lands or in countries engaged in totalitarian war, the result is the same, and the thing which has been most horrifying in those cases is precisely the thing which they have had in common. We could almost provide the mathematical formula for the evil; for it reduces itself to contempt for the other man's personality, the belief that the enemy 'deserves' killing, the refusal to contemplate the case of the innocent thousands who may have suffered at Hiroshima. Those who can drive motors and move mountains and survey the stars, may still be capable of failing to imagine the case, failing to feel the suffering or to be moved to mercy. And the men who in the 19th century expressed misgivings concerning the future of democracies, feared just this phenomenon. They feared most of all that civilisation would become submerged under those very 'wars for righteousness' which have been so disastrous in our time.

What is more, it is apparent that we cannot cure the evil merely by becoming like this ourselves, merely by falling into the same sins. That is one of the reasons why we have to beware of 'wars for righteousness' and beware of forming a counter-system to communism which must end by becoming all too like the system of communism itself. Special laws necessarily come into operation as soon as it is conceived that atrocity and cruelty are generalised, so that they represent a mass phenomenon; for once this has happened it means that something has gone wrong with the conditions of human life, and in a certain sense moral indignation is beside the point. It means that we cannot rectify the case by attacking the evil direct, but must address ourselves to the remedying of the conditions which produced the evil. In a similar way, special laws necessarily come into operation as soon as it is conceived that we are a civilised world which is now compelled to fight a type of barbarism; for at the root of barbarism there is something that is not disposed of by a moral judgment, and, though it may be necessary to be most competent in one's defensive measures, here is just the one great evil that can never be cured by mere punishment.

There was a time when Whigs and Tories thought equally badly of one another – in their great duel of Right *versus* Wrong. The executions procured by the one party when it was in the ascendant provoked the desire for reprisal in the other party when the situation was reversed. It would have been possible to go on indefinitely with such a conflict, and at every new turn of fortune the policy would have seemed more justifiable than before, since the enemy would have given more definite reason for punishment. In fact it is difficult to see any escape from such a circle unless somebody who has the power to inflict punishment chooses to withhold it, though the enemy seems to deserve it more than ever. The English political tradition was produced in the very process by which English party politics came to be turned into a form of 'limited warfare'. It was produced by virtue of the Christian interpretation of righteousness and by means of those very procedures which it is our present purpose to examine after they have been transposed on to the scale of international conflict. England is indeed the happiest of all the illustrations of the fact that civilisation itself requires the doctrine of the foregivness of sins. And the pattern of this doctrine reproduces itself in incredible manifoldness in the very cells that go to make the fabric of an urbane and tolerant world.

Note

1. Orginally published in *Christianity, Diplomacy and War* (New York, 1953), pp. 41–51.

10
Morality and Human Progress[1]

Moralism and the scientific approach

It is possibly a mistake to discuss politics in a separate intellectual realm which has principles of its own and in which the basic ideas, such as that of 'the state', can be puffed up into philosophical concepts. I think we ought to live and make our judgments as though in the mundane realm nothing but human beings matter, nothing but human beings exist; and if we use words like 'state' or 'society' we ought to avoid deifying abstract or collective nouns – accepting the terms as a kind of shorthand, perhaps, but not thinking of them as the pattern of something laid up in heaven – in fact, rather bearing in mind that behind them are just so many people. When I hear it asserted, therefore, that there is a separate ethic for the statesman, a peculiar thing called political morality, I am not sure that I can fit this into my thinking or even quite understand what it means. The statesman, like the professor or the poet, will constantly be confronted by the alternative between an act that is more moral and another that is less moral; but we can hardly admit of any fundamental difference between the quality of the decision in these cases, or the ethical principles involved. And in particular I don't see why in politics the virtues which I associate with the Christian religion should be suspended – humility, charity, self-judgment, and acceptance of the problem Providence sets before one; also a disposition not to direct affairs as a sovereign will in the world but to make one's action a form of cooperation with Providence.

It may be relevant, therefore, to stress the importance of that doctrine of individualism on which modern liberal politics has been so largely based. I don't agree with some Christians who seem to despise the doctrine, though I understand them if they say that it is a diminished and diluted version of an earlier and higher doctrine of personality. I differ from many non-Christians because, though they can go on holding the doctrine after having thrown overboard the Christian dogmas, I think that they inherited it from religion and hold it in their capacity as 'lapsed Christians', weakening it perhaps, and making it more precarious, by reason of the fact that they have cut it off from its source. In its profoundest sense, the doctrine emanates from our religious history, nor in fact would one ever have expected it to emanate from any other quarter, because religion had so great a part in deciding the way in which people stage the whole drama of human life on the earth. The more religion came to be related to the inner man, the more it had the effect of elevating the idea of personality as such, providing the individual with something which overrode the claims of mundane society and gave a basis for resisting the tyranny of mere majorities. The significant point in history, as Acton showed, was that it is the human being, not society that has the soul, the conscience, the cross-reference with eternity. From a mundane point of view this means that the individual has the right to choose the God Whom he will worship, and the moral end which he will serve.

On this view, men can not be written off as mere willful bundles of caprice or chance combinations of atoms. And they move to a diversity of ends – they are not herded together like cattle, harnessed and organised in a slave-system, and working for one general purpose which shall be the same for all, a purpose dictated to everybody by society. Progress is measured, not for example by the advance that is made in the exploitation of the resources of nature, but by the way in which human personalities are adding to their stature and the inner man is being enriched. And, fundamental to this kind of society, is respect for the other man's personality, the other man's end – in fact, the ability to put oneself in his place and see that he may have a conscience too. Secular thinking which starts from 'the rights of man' seems to lead to egotism, and has even produced the assertion that an individual need not obey a law unless he agrees with it. Those who pick up the other end of the stick and start with 'the duties of man' seem to end by putting the individual at the mercy of the state, which emerges as the super-personality. There are difficulties if human beings are regarded as absolute ends or as mere means to another end. That is why it is useful to think of both the individual and society as existing for the glory of God. Secular opinion itself must find something which transcends both – something which is higher than either man or the state and which provides a regulative principle.

Parliamentary government has not invariably succeeded; and, with young democracies in particular, there tends to emerge a certain willfulness of human nature that can lead to aberration. It seems that it is not sufficient to settle a satisfactory form of government on paper; certain 'imponderables', certain human factors, are necessary if the machinery is to work at all; no machine is proof against an absolute determination to misuse it. Both as an historian and as a contemporary observer, I am particularly interested in these 'imponderables' without which neither a democracy nor a League of Nations nor UNO seems able to produce the desired results. If in the Anglo-Saxon world there has been the necessary amount of the spirit of give-and-take, the disposition to compromise, the respect for the other man's opinion and the reluctance to resort to desperation policies, I do not know how far this may have been ultimately due to the survival of religious influences, or the effect of a long political experience, or the state of relative security, or a growing tradition of urbanity in a region comparatively free from violence. It is clear that democratic politics do require an internal discipline and a degree of self-control. Either one has to have a certain last *arrière-pensée* to the effect that one must allow for the chance of the other man being right after all, or one has to feel that by too great a fanaticism in respect of the immediate issue one may be doing harm to a higher object still, or one has to realise that even though one may be in the majority at a given moment, one must not provoke the other side by too great an exercise of power. In fact, the lack of the realisation that there may still be an object higher than the one that is the subject of dispute at the present moment is the thing which makes doctrinaire politics ever-arrogant on occasion; and it appears that those who have no religion are particularly liable to bring a religious fanaticism to problems of mundane organisation which ought to be matters for transaction and negotiation. Lord Acton was probably right when he said that liberty is impossible except among people who have a sense that the whole political game is being played in a realm over which there rules a higher law. In fact, a liberal régime cannot exist merely on the strength of my assertion of my own right to liberty. It depends on my saying to myself: 'This other man is like me and he must have the same degree of freedom that I claim for myself.'

All political action clearly has to stand under the judgment of moral principles, but for the sake of those principles themselves it is necessary to avoid confusing the moral with the merely moralistic. One can make a terrible miscalculation in war, and then submerge the whole issue in moral indignation against the enemy who has merely taken advantage of one's error. A careless librarian, who establishes no regular system for the checking of his books, may be satisfied just to heap blame on the people whose delinquencies have resulted in gaps in his shelves. We are on safe ground if we regard morality as the thing which we have to bring home to our own side, to our own nation, and in the last resort to ourselves. For the rest, where other people are in question, the real achievement is to induce them to perform an act of self-judgment in their turn.

But in respect of the standard which we set before ourselves, it would be proper to take note of the tremendous moral conflict that is recorded in the Gospels, the conflict conducted by our Lord not against the publicans and sinners but against the upright people, the most priggish moralisers of their day. Here, at the very heart of Christian teaching and Christian narrative, we see a Christ Who is reproved for consorting with the disreputable and Who stands in high combat with the righteousness of the stiff-necked. It seems clear to me that He does not judge on the superficial act, but proportions praise and blame to men's opportunities, to the materials they had to work on, and to the degree of temptation and test. The very name of Tyre and

Sidon had become a symbol of wickedness; but to the more respectable cities of His own day Christ says that if Tyre and Sidon had had the Son of God in their midst – they would have been better than modern Capernaum, and would have repented long ago in sackcloth and ashes. It is as though Christ were to say to Britain and America at the present day: 'Don't pride yourselves on being better than the Germans. If the Germans had been as fortunate as you have been, and if they had been in the same position, they might have behaved at least as virtuously as you.' This whole view – this withholding of the absolute condemnation of human beings themselves even where the actions as such call for condemnation – has received a high development in our own civilised order. One consequence of it has been the fact that conduct for which human beings were once blindly condemned has been found amenable at times to scientific method and even medical treatment.

Once we ourselves have abandoned Pharisaism and self-righteousness, the whole state of the question – in so far as it is a question presented to us – has to suffer a radical transposition. We ourselves are imperfect parts of an imperfectly ethical world and perhaps we need an Einstein to work out the ultimate consequences of this; for just as Newtonian principles stand firm in their own sphere but are englobed in a higher synthesis that reduces them to relativity, so in morality there seems to be a higher Absolute to which our absolutes have reference. We can judge ourselves but we cannot seriously make the absolute judgment on other men, because we do not know how to work out the calculations and the transpositions required. We can condemn an action and try to stop it if it is harmful, without even always feeling sure that we would never have committed it ourselves. The condemnation of other human beings is particularly questionable in that political realm in which our judgments are so gravely entangled with our interests. Our problem in fact can only be brought under control if at the beginning of the argument we see ourselves involved in man's universal sin; just as the problems of physics can be made amenable to mathematics when, instead of assuming that some bodies posses weight while others have the positive virtue of lightness or levity which makes them thrust upwards, one begins to see all bodies as simply having weight to a more or less degree. It must be particularly true in the political realm, furthermore, that the possibility of our doing anything to improve the world is greatly increased if we realise that we may even be responsible to a certain degree for other people's sins. Those who say that there is a sort of abstract Justice that they are called upon to vindicate, cost what it may to the world, are making the assumption: 'We are the righteous ones', and ignore the amount of wickedness and cruelty that they will still have to leave unpunished. Christian duty is not fully defined by the prohibition of specific forms of misconduct, and all these are englobed in a higher positive principle, which demands the unlimited and requires us to go as far as love can take us. And, even if we cannot always love, there is no hope of progress unless we base our calmer thinking on the truth that in an imperfectly ethical world it is love, not justice, that must be the criterion. For this reason, if we are considering a peace-treaty, we are not entitled to adopt any measure on the strength of a purely punitive argument, we are allowed only to look at the whole situation and ask: 'All things considered, what now is best for the world?'

Even if all this were not accepted as a general rule, it represents the attitude that we have to adopt when we are confronted by the problems of statesmanship. At a time when Gladstone was preparing for Disestablishment in Ireland, Queen Victoria sought to deter him by drawing his attention to the agrarian outrages that were taking place in that country, outrages which in her opinion merely called for condemnation and punishment. Certainly it is not necessary for us to approve of those outrages, and we may accept the view that the Irish peasantry had been committing sins and crimes which one would expect a religious confessor to reprove, the police to repress, and the courts of justice to condemn. Gladstone replied to the Queen, however, that the government was not surprised to see the outrages, that indeed such things were rather to be expected, human nature being what it is. This particular kind of crime in Ireland, he said, had something of the nature of a disease. The government was proposing to remedy the conditions which had helped to produce it, though it did not reckon that the remedy would come very quickly into effect.

I need not say that in the 20th century, as in previous ages – in foreign policy, as in internal affairs – we find here an almost unlimited field for pure arbitrariness. When demonstrations and rebellions are directed against us, they may be condemned as acts of criminal violence; but when they are directed against our enemies we may have eyes only for the criminal conditions which provoked them. And the case for Gladstone's attitude against that of Queen Victoria would be stronger still if the Whig historians of liberty were correct in their traditional assumption that, since tyranny pays little attention to reasonable entreaty or merely verbal protest, the only way in which people can show their discontent – the only way in which they can convince the world that something is hurting them – is by a desperate resort to violence. George III was wilful when he said that the American colonists were wicked to rebel and that there must be no negotiation until they had ceased the resort to arms. The truth was that, if they had not rebelled, the British government would never have believed that there was anything seriously wrong. If, however, we are not content to beat the air with mere shouts of condemnation, one way or another, it is sometimes possible to remedy conditions in such a way as to alter, not perhaps the essential heart of man, but the amount of violence and crime in the world. And Gladstone's determination to transcend mere Pharisaism particularly concerns us when we are faced not just with agrarian outrages in Ireland but with the problem of steering the world out of an era of wars and revolutions and apparently inescapable deadlocks. We shrink from applying the same principle to foreign policy and enquiring into the transpositions that this would necessitate.

If there were to be a great revolution in France tomorrow, I should expect that (apart from some unusual combination of virtue, skill and good fortune) it would be accompanied by terrible atrocities. I should expect also a dictatorship, and I do not agree with those people who think that the dictatorship has any organic connection with the ends that the revolutionaries are seeking to achieve. It is not to be linked to the Puritanism of Oliver Cromwell, and men were wrong when they thought that democracy as such was responsible for the Jacobin Terror during the French Revolution. Dictatorship is not even inherent in the ideal of the classless state; but it is absolutely germane to the modern revolutionary process – it has its origin and opportunity in revolution itself. The real menace is the phenomenon of revolution and perhaps, therefore, we have romanticised about this too much and wrongly idealised the conspiratorial type of politics.

Even subtle changes in society may have a significant effect on human conduct. An English businessman said some time ago that the complexity of the regulations nowadays and the disastrous results of the delays caused by red tape, had driven him to resort to practices which during the whole of his previous life he would have regarded as immoral. In periods of great upheaval the transformation of human nature may be greatly magnified; and men who have maintained all their lives a respectable conduct in an ordered society will sometimes seem transformed into brutes amid the terrors and the cut-throat struggles of a revolution. Even short of that, the political world may fall into such malady that everything is melodrama and the moral order seems to have collapsed, as in the days of Catherine de'Medici. Treachery abounds; but the suspicion of treachery is more abundant still; and, because you so suspect and fear the other party, your own behaviour is corrupted – the mere suspicion of treachery seems to breed more of the thing itself. In these circumstances, men considered virtuous in their private lives are sometimes guilty of great crimes in their public conduct; and in any case the conditions are such that the wrong men will tend to get to the top in politics.

But when at length normal times and stable conditions are restored, so that political life itself takes a more respectable and regular course, it is inadequate to say that men are being born better now, or that human nature has made itself superior to conditions, or even that better men are now coming to the top in public life. A whole group of imponderable factors are set in operation, and it is these which change men in general and secure a transformation of political life. Yet in a sense human nature itself is not basically changed; we have ample evidence that all the evils will return if the order of society breaks down again. A tremendous effect is produced on human conduct when desperate fear is removed; we see it on a small scale when a situation is brought about in which the man in possession of power no longer needs to feel that he is in danger of being executed if ever he falls from office. In the security of more stable times men begin to work, not just catch-as-catch-can, but with long-term purposes in mind and with a desire to facilitate the transaction of business. There have been times when it has been seen that even in diplomacy lying defeats its own ends and merely obstructs negotiation, so that an untruthful diplomat has appeared as a marked man. Then, also, something like a 'credit-system' develops, something as subtle and as significant as 'confidence' in the financial world – putting an end to your feeling that if you are as virtuous as you want to be you will merely make yourself the dupe of all the bad faith that exists around you. I think I could sum the matter up in the thesis that whatever credit may go to the social order for deterring the would-be murderer by the threat of punishment, still more must be due to it for the production of a stability and a mode of life in which it does not occur to the vast majority of men to want to commit murder, nor does it occur to the majority of men even to fear such a fate. All this explains why I personally would attach more importance than most people do to the ancient view that the object of political society is to establish an area of peace and order in which men can develop the life of reason and the world can grow in reasonableness. And I think that the state, without removing human sinfulness, is the means of establishing a tolerable mode of life, establishing a field for the play of reason, in spite of sin. In one sense it is the remedy for sin; because at least the external conduct of men becomes controlled, and the reign of actual violence is curbed. All the 'imponderables' which are set in operation in a stable society have a similar relevance in respect of the existence also of an international order. Here, too, mere 'prohibition' is ineffective except when there is a state of things which induces also 'inhibition'.

To take now the extreme case: the occurrence of multiple atrocities, especially on the part of governments or great masses of people, is so associated with backward states of society, revolutionary situations, social breakdowns, and the cataclysm of war, that, though the culprits might be morally indefensible, it is a question whether statesmanship should not confront the problem as rather an example of the breakdown of civilisation – a 'reversion of barbarism'. It seems clear that under certain conditions a society can lose for example a great deal of that respect for human life which means so much in a civilised world. Here – at the very point where we might find that we ought to bring moral indignation to its climax – we must wonder whether we are not confronted with the supreme example of the kind of evil which, in the ordinary way, can only be rectified by remedying conditions. If this is true, here is a case in which statesmanship can do no more than cooperate with the historical process; for when a moral order has collapsed in this way, the remedy must be partly the work of Time.

It follows from all this – or it is inherent in all this – that, while holding what I think must be extreme views concerning the play of personality and contingency in history – concerning the uniqueness of all the successive combinations of circumstance and the impossibility of predicting future turns of events – I consider that these things, and even the whole of history, have a vast substratum which is amenable to what might be called 'scientific' treatment, and at least provides us with a way of achieving more intellectual control. It seems to me that fifty years ago history and natural science confronted one another as something like equal powers; and if scientific thinking has marched ahead of historical thinking – if our control over nature has outrun our science of human relations – I wonder whether it is not because of the audacity with which the scientists confronted (instead of evading) the intellectual paradoxes which kept emerging from this work. Assuming that the War of 1914 might have been avoided if one or two men had made the different decisions that it was open to them to make, it is still important to relate those decisions to conditioning circumstances (such as the constitution of Germany and the state of the Habsburg Empire) and to see the cataclysm in the light of an historical process traceable for forty or perhaps a hundred years. Statesmen can in fact say that, 'human nature being what it is' such-and-such a piece of political high-handedness is calculated to provoke people to a desperate act of violence; and this is one stage wiser than being politically high-handed and then indulging in moral indignation against enraged men after they have resorted to force. Gladstone was not lacking in respect for personality when, as against Queen Victoria, he saw the need for love rather than judgment – addressing himself to the problem of conditions, even though it might entail treating Irish peasants as not superior to conditions, not quite masters of themselves.

In fact, maxims of government, and reflections on history as a process, are as ancient as the endeavor to discover correspondences in the inanimate world. And the question is whether the thinking may not be carried further than is customary at present. Perhaps the crudest kind of example which one could quote – and I may be expressing it with the maximum crudeness – is the thesis that revolutions are subject to a process which makes them tend to move ever further to the left, or to greater extremes. There is certainly some evidence that might lead one to question the thesis (which must complicate any moral judgment we try to make on the victims of this process) but it is possible that at a further remove this evidence only serves to strengthen the argument. Where men have successfully checked the tendency that is here in question, this seems to have been because they had been so highly conscious of the tendency in the first place, so that they took special measures to counter-act it. The very knowledge of the law or the generalisation or the process or the tendency may provoke the action which enables one to falsify the thesis – to prevail over the very tendency that one has diagnosed. But we must not imagine that the value of the 'science' is nullified merely because the 'science' has enabled us to bring the whole matter under intellectual control. The thesis that revolutions move to the left is a flexible one, therefore (because once we are aware of it we can do something about it); but it is by no means useless on that account. If we say that it is an historical law, it is one which bends so to speak – bends but does not necessarily break. At the end of everything, it is a law which no wise leader of a revolution will ignore. The very examples which contradict it do not necessarily make it invalid; and we have to say rather that they force us to a higher altitude – force us to subsume the original generalisation in something higher still.

Allowing for what is unpredictable in the wills of individual people and what is unique in every historical conjuncture, something is within our reach which might almost be called a 'science,' if we can conceive of a 'science' which can dominate a vast area of its subject matter (i.e. the political conduct of men), though it still always has to travel, so to speak, with an unknown quantity, and its results, therefore, need to be handled with flexibility. The laws which are produced are perhaps true in a conditional sense only – true sometimes only 'other things being equal,' or valid only as representing a tendency or a bias in events that we must take account of. Or they may stand as formulas that provide what might be called a norm; and it does not necessarily matter if in every actual historical case there may be some different kind of departure from the norm. Similarly there may be generalisations about people in the mass–even formulations of a process which takes place in the whole mass – yet any single individual may seem to contravene the rule; and where there is an exception to the whole generalisation, this exception may not affect the argument, or may only cause us to become aware of a new factor that drives our thinking higher. It seems to me that all these are the kind of paradoxes which the natural scientists of the present day are not allowing to defeat them; but the historian, who met them first and ought to have resolved them first, is allowing himself to be checked too easily. Ours is a realm in which purely mechanical methods may be fallacious and a wooden literal-mindedness has sometimes been fatal; and what I am describing is perhaps rather the operation of the mind of the statesman, who has ranged over past and present, has reflected on the results of his own actions, has hunted out the correspondences in events, and, with the help of insight, has reflected methodically on the total results of 'experience'. In all this, individual human beings are being treated as more highly conditioned by history and necessity than *prima facie* is often assumed. But till we know the necessity which limits us, we cannot come to understand our freedom either. The very fact that we analyse our position more profoundly and handle it more scientifically, means that we are able to steal a further march on events and subdue a wider area to our intellect and will.

It entails what I have described as the morality of the Christian rather than the moralism of the Pharisee; for it is not satisfied with the judgment that the war of 1914 or the atrocities of the Jacobin dictatorship are the work of particularly wicked men, and it insists on a further analysis that reaches behind the conduct so condemned. The moralist and the teacher, the prophet and the preacher, address themselves to the improvement of human nature itself and may be justified in working through direct exhortation and judgment. The statesman is concerned to improve human conduct rather by the process of rectifying conditions.

It is for this reason that international affairs are not merely a field for a statesman's improvisation but also a region for what might be called scientific investigation. And the statesman, compelled to improvise as particular conjunctures arose, is better fitted for even this task if, without losing his flexibility of mind, he takes his decisions against a background of ideas that are the result of scientific analysis – including a proper notion of the larger historical processes with which he is confronted and which should affect his long-term conceptions of policy.

Foreign policy and historical processes

The concept of *raison d'état* has had a significant part to play in the development of an international order. At the relevant period – in the time of Richelieu, for example – it was a valuable instrument, for it helped to carry the world over a necessary transition. Not only subjects but kings themselves had to learn that authority must rise above the play of private interests and above purely dynastic considerations. What was needed was the notion of a public policy which should envisage the welfare of the realm or the nation as a whole. But if kings and nobles may be merely partisan or merely self-regarding, there is also a selfishness that men may have on behalf of the body politic, and it may seem a deceptively public-spirited thing. Once the transition to the modern idea of the state had actually been accomplished, *raison d'état* could only survive as a synonym for national egotism, a pretext for corporate covetousness. Even in the 18th century, the more serious reflection on the subject of foreign policy assumed that the preservation of the international order was an overriding objective for the practical statesman. Even if in the last resort it was based on a prudential calculation, there was some recognition of the fact that policy is referable to an objective higher than the state itself.

It can hardly be denied that the foreign offices of the world do set out primarily to take care of the interests of their own country; they are in fact, and they conceive themselves to be, subject to the pressure of interests inside the country itself. They would fear the reproaches of their own public if their policy had the effect of lowering the standard of living at home, or undermining security, or weakening the voice of their nation in world affairs. It is possible, at least under certain conditions – and it would be invaluable to know the conditions – for states to behave with no more than a reasonable and conventional amount of egotism. Some may appear altruistic and put out the most moralistic protestations; and in the absence of earnest self-analysis, they may fail to realise how even these can be the cover for aggressiveness and vindictiveness. On occasion the benevolence may be genuine and the altruism deeply felt, and yet one may be unconscious of the fact that for the time being virtue has been made easy – that peace, for example, is particularly important for one's own interests. All the same, a state may fairly acquire virtue from the very fact that it contrives to make its self-interest harmonise with something that is good for the world in general. I am not clear that any state ever surrenders empire voluntarily, though wise ones have been willing in the last resort to give way after merely measuring forces, anticipating the result of a possible resort to war. I do not know that any state would surrender an imperial position in the world if it thought it could keep such a position by taking to arms; though it is possible that a people intent on private purposes and weak in public spirit would allow their power as a nation to wither away.

On moral grounds, as well as on prudential calculations, national egotism requires to be checked, superseded and transcended. The most dangerous argument that can be used for justifying such egotism is the assertion that since my own country wants to work for righteousness, or my own country has the right idea of what is virtuous; its power to do good in the world must be enhanced as far as possible. In other words, our own side must not be allowed to be weakened because it is our side that represents the good; and if we even lose a strategic position, or neglect to acquire one, we can convince ourselves that this must be the betrayal of a cause. Yet if we can use this argument our enemy can use it, for it is an argument available to every nation in the world. Only God can know which is in the right; for in reality each is acting as the judge in its own cause. Here, in fact, is the essential formula for that intellectual *impasse* with which any serious thought on our international problems must find itself engaged. If a party within a country (or a national government in its relations with the world at large) is so strong in its sense of injustice or its zeal for a certain kind of righteousness that it will not give way at any price, it may convince itself that it is putting morality in the supreme position. But is is egotistical in claiming the right to stand as the highest judge, and in reality it is only saying that the issue shall be decided by war.

We may imagine that our position at the present day is unique; but in reality Europe, during the last few centuries, has had a considerable experience of creeds, ideologies and régimes whose claims were absolute – conflicting systems which were mutually exclusive. Catholicism, as the authentic form of Christianity, insisted that it possessed universal authority. The Protestants could not believe that if their religion was true any other could avoid being a diabolical blasphemy. The French Revolution was missionary in intention and there was a time when democracy proclaimed something like a war of religion against kings – until gradually its ardor was tamed and it turned over to something more like conventional diplomacy. All such creeds and ideologies and règimes, of course, would have conquered the world by force if they could, and would conquer it still, at the present day, I think, if they saw the chance. When people think themselves so profoundly in the right they would feel that they were delinquent if they declined to promote their cause by actual fighting. And this is the case if the creed is that of gentleness or the programme is that of liberalism – it seems even to have been the case with people whose programme was the abolition of war itself. Single nations would have achieved 'universal dominion' if they could; and if France ever ceased to be the 'eternal aggressor', as we in England used to call her, this, too, was not due to unconditioned choice, or an act of charity, or a change in the fundamental nature of the Frenchman. Whether a nation or a creed or an ideology or a regime is in question, nothing useful can begin to happen in the international world until the aggression is checked.

But this does not mean that we are permitted to enter on a course of counter-aggression, and if we say that we will wipe out the offender we are committing the very offence against which we had been driven to fight. Almost all the great aggressors of modern history would seem to have taken to conquest by the indefinite extension of what had originally been counter-aggression. When human beings move to such a diversity of ends – some of them crying out for Catholicism alone, some for Calvinism alone, some for democracy alone, and some for nothing short of the classless state – the function of diplomacy is prescribed by the very nature of things. If a foreign office wishes to aim at higher things than mere national egotism – even if it is merely far-sighted in its selfishness – and supposing it can do more than merely skirmish among present dangers, desperately trying to keep head above water – it has to set itself a more specialised objective. The aim is certainly 'the good of mankind', but this vague entity has to come down to earth – it condenses itself into something concrete and well-defined. Precisely because there are nations working for their exclusive interests, or ideals and ideologies making their absolute claims, the real object of foreign policy is the establishment of a *modus vivendi*. The work is a creative and original thing; for it implies the discovery of a *modus vivendi* for bodies and systems which at the beginning of the argument are incapable by definition of any common existence. When the object has been secured, it is like religious toleration for the Protestants and Catholics; one comes to think of it as part of the natural order and feels that men must have been willful not to have established it earlier. But whatever it is, the *modus vivendi*, once it has been established, implies that the fate of all the other purposes and creeds can be left to the play of human reason and the operation of opinion; it does not depend any longer on the mere hazard of war. If we are devoted to the idea of a liberal world in which men shall be Catholic or Protestant, and regimes shall be monarchical or democratic, according to necessity or desire, the argument for taking this way out of the *impasse* would appear to be more cogent than ever. In the international world it is peace that is the primary object; and this means the development of an international order in which the issue can be decided by reason and the world can grow in reasonableness. In fact, the higher the cause – and no matter if the cause is that of Christianity itself – the more urgent is the necessity for extracting it from among the purposes of diplomacy or the objectives of the belligerent powers. The existence of an international order means that we can rescue the loftier ideals from the dominion of force and the play of mere mischance.

Under this system, whether we are practising diplomacy, or conducting a war, or negotiating a peace treaty, our ultimate objective is the maintenance and the development of an international order. This is the purpose which transcends national egotism and puts the boundary to self-interest – the purpose to which all our more immediate aims in foreign policy have reference. Even if one of the members of an inter-related group of powers is less civilised or more barbarous than the others, it is possible that by comprising it in the international order – by making it a member of the club – we are taking the shortest way to securing its moral progress. It might be objected that there exists some power in the world which is determined not to submit to the rules of international order. Such a consideration so underestimates the real predicament, however, that it may be said to misconceive the whole problem. It is important that we should confront this particular issue at its point of maximum intensity.

Let us remember that no single power, no single creed or regime, is ever for a moment prepared to submit to the rules at the beginning of the story. All of them have objectives which they claim to be more urgent than peace, objectives which they can claim to be higher than that of mere international order as such. Deadlock would in fact be absolute, if one left out of account the work of time and the historical process. The truth is that men may differ greatly from one another in their predominant end or their chief preoccupation. For one it may be religion, for another socialism, for another music, for another business, for another family. The operation of time and human management, however, may bring them into curious unanimity about the object that comes next. Once their aggression is checked, they want peace; and if they cannot have the exclusive domination of the world they are determined that none of their rivals shall do so either. In fact, if they cease exultantly conquering, they prefer that the issue shall be left to the play of reason. Those who believe in their own cause will tell themselves that they are bound to win if only they can have peace and fair play – if only the issue can be decided rationally. Precisely because of their faith, they have the idea that in the last resort the historical process is on their side. In the long run everybody at least wants everybody else to submit to the rules of a civilised world.

In any case, it is not all people who can fasten their minds on one objective, keeping the elastic at stretch all the time, sacrificing the good of their family, their trade and industry, their standard of living – forgetting about art, amusement and falling in love. Governments themselves, if they see that victory is receding, will come in time to set a higher value on mere security. They discover that everything is being kept at strain, that they are trying the devotion of the subjects, and that they risk the loss of what has already been achieved. They change their objectives, and some of the men who once gloried in a war for the principle of self-determination have come to find that in a new context they are not so sure of their devotion to that principle. Even if a people are in a state of frenzy, every year in which they taste security makes them love security more, especially as they settle into the preoccupations of ordinary life. And, for the sake of mere peace and order, nations in the past – and even incipient democracies, jaded with tumult – have surrendered to a kind of dictatorship. It is not as though the acceptance of an international order entailed the loss of anything fundamental in their higher ideals. They sacrifice nothing at the finish save the claim to achieve universality for those ideals by the use of force. The religious parties of the 16th century abandoned those principles that made them feel obligated to use force or to scorn the rules of social order. Neither Catholic nor Protestant, today, however, feels that it has surrendered anything essential to itself – surrendered even the claim to exclusive truth. In the end the greatest paradox of all takes place – each of the opposed systems tries to borrow the best from the other, Protestants being influenced by Catholicism, Catholics being influenced by liberal ideals, monarchical statesmen like Bismarck taking over some of the ideas of the French Revolution.

The whole process which I have been describing is so common in history that I wonder whether I ought not to call it inevitable; and indeed if something of this kind were not implicit in the workings of history, it is not clear how the world could ever have developed from a state of barbarism to a civilised condition. The wilfulness of a statesman or an unlucky combination of circumstances may still provoke a war or an upheaval that puts everything back to its original state, but it is a mistake to overlook the healing effects of time; and statesmanship, though it may hinder the process, may come into alliance with it, and greatly add to its speed. I want to focus attention on the process because it affects our conception of the kind of good we think we can achieve in the world by politics and diplomacy. The issue is: Do we fight a cosmic war of righteousness, Protestant versus Catholic, non-communist versus communist? Or, are we fighting for a world in which Protestant and Catholic, communist and non-communist, achieve a *modus vivendi*, the actual conflict being relegated to the realm of rational discussion and the play of opinion? This is one of the points at which I seem to differ from other people; for I think it best not to be pharisaical in politics – best to ask whether we ourselves may not be wrong somewhere, or partly in the wrong, or partly the cause of other people being in the wrong. And I think we should be straining to put ourselves in the position of the other party, trying to see his case at its best – trying to put the enemy's case better even than he is able to put it for himself, and without those blemishes that might be merely incidental to it. In other words, even in foreign policy, I don't see why (at the highest level of our thinking at least) the theological virtues of humility and charity should be suspended. And the real negation of both politics and diplomacy (as well as religion) is to say that our view shall prevail, whatever it costs the world.

Wars, revolutions, states of absolute deadlock – these are things that the world has repeatedly been confronted with, as far back as history can see. The crucial problem that we have to meet today – the crown of all problems in politics and diplomacy – is precisely the ancient one: how to steer the world out of an era of war, out of a revolutionary situation, out of the absolute *impasse*. It all looks impossible, but history is full of admirable laboratory specimens showing how at least it has been done in parallel situations throughout the centuries – how the thing which in logic seemed absolute deadlock can in fact be circumvented. But the cooperation of time and the historical process are indispensable. That is why it is such a terrible thing to carry war to such a point that when hostilities are over it is impossible to recover the international order – the sense of all being members of the same club.

It might properly be argued that, over and above any controversy concerning creeds, ideologies and regimes, there is the problem of a government which commits atrocities at home—barbarities the wickedness of which cannot be dissolved in any sort of relativism – inhumanities which challenge our indifference, justifying a 'war for righteousness'. Certainly it is true that the severities imposed on people by governments – and particularly by revolutionary governments – as well as those imposed for example by a certain racial or national group, may be of a particularly revolting kind. Even when our diplomacy has been ideological, however, we have not in recent decades regarded such things as a reason for actual war; and though we may take them into account once war has broken out for other reasons, it is clear that they have not prevented attempts to secure regular diplomatic intercourse with the offending country up to the moment when such a war has occurred. I imagine that the principle of 'abstention' has come to be adopted in the world because it represented the lesser evil – on the long-term view it promoted the good of mankind. The object, I suppose, is to see that the evil comes to an end as soon as possible, rather than to secure its punishment, and war itself would seem to be calculated to increase the area of the evil and the suffering. As the world is at present organised, part of the work of civilising and humanising it has to be focused to a great degree on the smaller area of the nation. While this state of things exists, the resort to war for the purpose of rectifying the internal government of another country would be calculated to multiply the pretexts for war in every part of the globe; for there are few countries that would not provide a reason for intervention in one respect or another. In any case, it is easy to rail at sins which are being committed at the other end of the world, when the situation is such that the cause of civilisation might be better served if everybody concentrated on the abuses in his own country and on raising the whole level of life at home. In regard to abuses and atrocities, men have an ability to localise their vision, to see the sins of their enemies, to put the soft-pedal on exactly the same sins on the part of their allies, and, when a nation has become hostile, to remember at last the sins that were ignored while it was on their side. It is possibly even true that one is in a better position to modify the brutalities of a more barbarian country if one does not send that state to Coventry or refuse all intercourse; and I do not know whether England could not hope to influence South Africa more by keeping her in the Commonwealth than by declaring that she was not fit to be a member of it. I do not know whether there would not be a limit to 'non-intervention'; but I take it that the formula which justifies such abstention is the one which also puts limits to it: namely, that it is our duty to remove evil and hardship up to the point at which our intervention would increase the evil in the world and produce an aggravation of the hardship. Among the possible evils of intervention by force in another country's internal affairs would be the fact that the principle would become generalised and is particularly liable to be abused. And in so far as there is question of one country being more 'barbarous' and less civilised than its neighbors, it is not ordinarily to be expected that war itself will hasten the progress of civilisation, or that victory itself would end the evil for a much longer period than the victor-powers kept the defeated country under their direct control. Possibly, as the world is at present organised, it is only possible to have an international order at all, provided the principle is adopted of not intervening by force in the internal government of other countries.

It might even be said that we can too easily find satisfaction in the fact that the enemy (or the potential enemy) of the moment is committing atrocities at home and is ruled by unscrupulous men. This may enable us to evade the real problem which is set for the human intellect – the problem presented by the structure of the existing situation. It may be useful to examine the position of Stalin in this respect – not the myriad things which were incidental to his peculiar case, but the example of a man who is in the general position of a Stalin. Only in this way can we really face the difficulties presented to us by the twin phenomena of Revolution and War. We may assume that our Stalin will be a ruthless man, but this particular aspect of him, important though it may be in other respects, requires no special explanation; ruthless men undoubtedly exist in all societies and may be regarded as always lying in wait for their opportunity. What requires note is the revolutionary situation which tends to bring such a man to the top, and, when he has reached the top, gives him more power than the Czars possessed, and more opportunity for drastic decisions. We might almost say that, apart from exceptionally fortunate conjunctures, Stalin's personal qualities are not likely to affect the main structure of our problem;

for if he had really been gentle or had shrunk from severe measures – if he had the disposition of a Kerensky – he could hardly have been expected to hold his power against men more ruthless than he. In any event, a Stalin, once he has attained his position of supremacy, is likely to be obsessed for a long time by the realisation of its precariousness; and prospect of a fall from power is terrible to a man in this situation – a man who will have committed many crimes that the world is waiting to avenge. If he deals remorselessly with his colleagues, we can hardly tell whether he is an inhuman tyrant, jealous of all men of ability, or a ruler in a revolutionary situation dominated by fear. Perhaps only God can know whether in any case he is a fanatical idealist, sacrificing himself and everybody else to his cause, or an unscrupulous exploiter of other people's ideals. It may even be true that a man in the position of a Stalin was earnest enough at first but then was corrupted by the possession of power. Alternatively, he may have been correct in his initial diagnosis of the danger of his regime, but allowed the fear and the emergency-situation to continue longer than a wise ruler need have done.

But the important point to note is that statesmanship does not depend on the answer that we give to these questions – questions which in reality have reference to the interior of a human being. The man in the position of a Stalin may be a fervent idealist, convinced of the wickedness of the rest of the world – but we still have to defend our homes, our values, our traditions, our whole order. If we could pierce into his heart and find that he was an idealist, dominated by a terrible fear, we could not allow him to acquire a single province the more. And even if we knew that he never would abuse the additional strength which this would procure for him, how could we be sure that next year he would not be succeeded by an unscrupulous man? At the first stage of the argument, what we require is a policy directed not with reference to particular human beings but rather to the structure of the situation itself. We have advanced one step further if we have brought home to ourselves the fact that the enemy is revolution as such, the menace is the revolutionised state, the problem is the continuance of what might be called the revolutionary situation. It is important that we should face this problem itself – the problem as it would exist if the rulers of Russia at the present moment were the best men that conditions in that country could bring to the top. The structural problem must be disentangled from incidentals and personalities. Then, at the second stage of the argument, we can see the qualifications that might have to be made in view of the personal qualities of the rulers of Russia, or the more casual combinations of circumstance. In any case, it is the former of these states or enquiries that ought to be envisaged by academic enquirers, whose work must have reference to long-term ideas rather than short-range policies.

It is possible that in the field of foreign policy itself – and particularly under stable conditions or in what we call 'normal' times – the play of personality and the operation of incidental factors complicates the whole process of events in a more drastic manner than in the case of the dictator emerging during a revolution. In times of great uprooting, however, when the international order has been gravely shaken, it is possible that events respond more immediately to the power-situation and to the operation of necessity. Whatever the play of personality and of the more incidental factors, however, it is surely wise even in the case of foreign policy to direct one's attention to the whole structure of the impersonal situation – including the disposition of forces in the world – at least at the first stage in the argument. It is at a later stage that we can make readjustments in order to allow for the things that are contingent; and we are evading our problem if we direct ourselves in the first place to the incidentals and the personal factors in the problem. On this view, the position of a Stalin at the head of a powerful Russia is subject to the same kind of analysis as the position of Stalin the dictator arising out of revolution. Indeed, one might say that if a powerful Russia (a Russia that would be frightening in the world of power-politics whatever its regime) is at the same time a revolutionised country, one must expect over-facing difficulties, unless there emerges a leader exceptionally virtuous and wise.

We must never forget the terrible recurring fear of the Russian bogy, even in the days of the ramshackle Czarist Empire. At a time when the international order was more firmly established and recognised than now, it was never doubted that Bourbon France and Czarist Russia, for example, would be a menace to the world if ever the power-situation gave them the opportunity. The revolution of 1917, like the revolution of 1789, has had the effect of greatly intensifying the motive for aggression, and at the same time enlarging the opportunity. Even so, it appears that the revolution tends to follow the same channels of polity and the same lines of aggression as the Monarchy which preceded it. Moreover, when monarchs had it in their power to expand their territory, they never seemed to lack the religious or ideological pretext and they were often not destitute of the requisite dynamic. If we ask what the Revolution adds to the power of a possible aggressor, it may be replied that it brings a missionary fervor, the pretext for interfering in the internal affairs of other countries, and the conviction that all other governments are mere abuses. Also it provides a greatly enhanced opportunity for organising the state into an effective instrument of war. The very fact that it issues in a dictatorship means that peace is more likely to be at the mercy of a single man. All the difficulties are intensified if, apart from the revolution, there has been an upheaval in the international order itself, so that this is no longer part of the accepted framework of life. This latter presents us with a situation in which, in any case, it is likely that frontiers will be continually questioned and certain provinces repeatedly changing hands. Everything is therefore more fluid, more open to competition, more sensitive to displacements of force. This is a situation in which even countries that have had no revolution are often tempted to indulge in adventurous policies.

It is possible that our attitude to the international problem would be affected if we realised that, behind the personal wickedness of individual rulers who confront us, there is a situation which has been operating in favor of aggression. We are suffering the double consequences of Revolution and War; and since the victories of communism from 1917 have been so closely associated with war, perhaps what we are really suffering from is the heritage of 1914. If it is true that communism in our time has been the chief beneficiary of war (and has even prevailed against those powers that were victors in war itself), we shall be skeptical of the possibility of destroying it by means of victory in war – indeed that defeat of Russia and China would not wipe it out. The healing processes of time can only really begin when we realise that they are going to have to live with us. Here, to a supreme degree we have the kind of situation in which, once the aggressor is held in check, and once a balance of forces is achieved, the healing processes of time, and these alone, can solve our problem – can even secure that in Russia there shall exist the conditions for more virtuous statesmanship or the conditions that will bring more virtuous men to the top. Never was there a kind of case in which it was more clearly the duty of the statesman to act as one who is trying to cooperate with the historical process itself; and even the short-range policies that might have to be improvised for desperate conjunctures ought to be considered with this long-term policy in the background. Never, therefore, was there a kind of case in which it was more necessary that the operations of statesmen should not be deflected by false doctrine, false analyses of the situation, false views about the ends of diplomacy itself. All this is an additional argument for the scientific attitude and for the establishment of academic discussion on these issues.

This reasoning must apply *a fortiori* if it is noted that there is one danger that particularly threatens a genuine democracy, and is particularly liable to obstruct the natural processes of *détente* once the time is ripe for these processes. Since the beginning of the twentieth century at least, governments have persuaded their peoples to arm by teaching them that there was one particular villain in the world worse than the rest of human nature – one particular villain against whom it might be required to wage not merely the old type of 'just war' but a 'war for righteousness'. If the democracies were to lose strength in the present period of transition the very conditions necessary for checking aggression and producing a *détente* would be removed. The problem is how to persuade a democracy to keep strong without the hysteria associated with absolute hatred and the 'war for righteousness' – how to persuade democracy that in politics even generosity is not generosity if it proceeds out of weakness. The seriousness of this problem makes one understand (but not necessarily approve) the policy of the Hohenzollern dynasty in asserting that the army should not be subject to the caprice (and possibly the sentimental snatch vote) of a parliamentary assembly. It seems to me that, whereas we have tended to assume that democracy is only fit for a simplified foreign policy – only capable of being galvanised by hysteria into a 'war for righteousness' – there are signs that it will follow wise leadership, and in any case its future depends on its realizing the need for strength (while the world is constituted as at present) without having to be driven to this by the manufacture of hysterias. Otherwise the authoritarian state is likely to have a serious advantage as against the democracies. I see no escape, therefore, from the view that democracies, instead of being pampered by simplified teaching which precludes the real practice of diplomacy, must be educated into a proper doctrine of international relations, however difficult this may seem. It emphasises the need for the establishment of a proper doctrine and for the analytical treatment of the ideas and processes which must underlie foreign policy. If it is true that the psychological *détente* must precede any diminution of military strength, and that the inversion of the order here (i.e. too early a diminution of strength) might be fatal, the argument for this becomes stronger still.

On the long-term view, then, the statesman has to cooperate with Providence and the historical process; and here is a further reason why we should study scientifically the process of things in time. Modern revolution and modern war are twin phenomena; they undermine that political order which men come to accept and take for granted as the framework for everything else in life; and then these very things, instead of regulating human passion, become themselves the object of unremitting fanaticism. War and revolution destroy the 'credit system', they abolish the automatic checks on the politics of mere adventure, on the disposition to gamble with human destiny. They halt that deeper process by which the world gradually grows in reasonableness. At whatever point a loosening of tension becomes feasible or begins to take place, the partisans of ideological diplomacy and ideological warfare will be compelled to eat their words. Those who believe in the possibility of an international order are misconceiving their own position today unless they see that it means the gradual separation of the doctrinal conflict from the operation of diplomacy as such. It means that our diplomatic object cannot be the annihilation of a doctrine, and has to be the creation of a world in which communist and non-communist have mutual intercourse. Our wiser course, therefore, is to direct our attack against revolution as such rather than against the objectives of a particular revolution. Our object is not even the counter-revolution; for it is not clear that the civilising process would be hastened by the advent of a new revolutionary system again in Russia – one which might have to carry the story back to the beginning. There is in fact a chance that all parties – revolutionary and non-revolutionary, each in their own countries – are going to develop into anxious defenders of their own *status quo*. Both sides will be striving to preserve their established order from revolutionary menaces and resistance movements – from the illegitimate resort to force, intrigue and chance which leads to the overthrow of the moral order itself. It is not the classless state as such that is the real enemy, particularly if the processes of history are only likely enough to turn it into something else. And if we accept this policy we are less in danger of covering or defending any abuses that may exist in our half of the world – in fact, we are rather asserting those blessings of freedom, order and continuity which outweigh all our defects.

For the time being, therefore, we may reach nothing better than the interim stage at which the prevailing principle is *cujus region ejus religio*, both Marxist and anti-Marxist making this concession to a principle of nationality and local sovereignty which the world has not yet found is possible to transcend. The universe is so constituted that it can bear the paradoxes of such a situation, if we ourselves will only accept them. Indeed it will force us to accept them in time, when we are weary of kicking against the pricks. The conflict between Catholicism and Protestantism is fertile in the production of those anomalous intermediate stages which illustrate the various patterns of a *modus vivendi*, and the various paths by which it has been reached in the past. Under the principle *cujus region ejus religio* one tolerated abroad the heresy that one repressed at home, and possibly one even made alliance with a state which had established that very doctrine as orthodoxy. At a further interim stage, Richelieu tolerated the Huguenots as a religious body, saying that he would count on their conversion by reason and example, though he ruthlessly destroyed them as a political power. In some countries *cujus region ejus religio* meant that the government, the monarch decided the religion of his subjects. In Sweden it was decided that it meant the rejection of a king who did not conform to the religion of his people. Richelieu is responsible, I think, for the conscious initiation of the idea of diplomacy and war conducted for limited purposes; for to rescue his country from danger he set himself against the Catholic Hapsburg dynasty, but, being an earnest Catholic himself, he used all his ingenuity to prevent the conflict from hurting Catholicism. Men have a way of making a virtue of necessity, and some of the compromises and adjustments that were taken unwillingly out of a desperate need to discover a *modus vivendi* – the idea of toleration itself, for example – later became ideals in their own right. The same technique may not be appropriate for the present day; but a diplomacy directed to its proper ends can be a creative thing, adding to the achievements of the human intellect, and producing devices for getting round an *impasse* that seemed insuperable. Here again, the scientific approach can succeed where mere moralism fails.

Note

1. The original manuscript of this chapter is in: BC File 31 (eds).

11
Moral Judgments in History[1]

I

Those who go to the works of Protestant and Catholic partisans will find the history of the Reformation staged as a conflict between right and wrong. If they do not become war-weary they will soon be somewhat bewildered; and they will yearn on occasion for a more humdrum narrative – one which will make clear the ascertained data and not harass the mind too severely by entangling the argument with the evidence. What they will hanker after, strangely enough – though they might be shocked to see the matter so expressed – is the technical historian who may be a Catholic or a Protestant but is willing to jettison for the time being his private views and personal valuations. He performs an act of self-emptying in order to seek the kind of truths which do not go further than the tangible evidence warrants, the kind of truths which the evidence forces us to believe whether we like them or not. Though this kind of history is more limited in some respects than a Protestant version of the story (which will be packed with Protestant evaluations and judgments) it is richer and more far-reaching in other ways, for it is calculated to carry the student to higher altitudes and it can lead him to a further range of discoveries. It may enable him, for example, to get behind the very conflict that divides Protestant and Catholic, and learn why Western Christendom ever came to such a state of dissension. This, then, is what we mean by 'technical history' – the sort of history which is the subject of a high and austere academic discipline. It may never exist in its absolute purity. But its assertions have a higher authenticity in so far as the ideal is attained.

Men who in centuries long past were firmly convinced that an outbreak of plague was the manifestation of the handiwork of God would still seem to have been capable of noting on occasion the observable connections between events and the operation of intermediate causes. Those who have believed that a war came as a judgment from Heaven have still been able to observe and discuss the activity of the human instruments of the divine Providence. A natural scientist is pledged to work in the way Gibbon purported to do; that is to say, he confines his explanations to the causes that are 'under God', and he would be committing an act of sabotage if he brought God into his scientific argument. It would be wrong to infer that because he submits to this discipline he is necessarily an unbeliever. The historian may be convinced that the will of God is in every step and motion of the drama, every pulse-beat of the centuries. As a technical student, however, he is under the obligation to perform the particular act of self-emptying that has been described. He is committed to an attempt to learn all that can be learned by the scientific study of just the observable interconnections of events.

If we consider these facts, if we note the nature of the questions which the technical procedure of the historian is intended and qualified to answer, if we bear in mind the intellectual realm in which that kind of question is appropriately discussed, we are in a position to embark upon an estimate of the place which moral judgments ought to have in history. And in view of the situation that has been described, it may be possible to reduce the shock sometimes produced by the thesis which denies any ethical character (in the usual sense of the words) to the technical historian's universe. The thesis may be asserted in so far as it means that moral judgments on human beings are by their nature irrelevant to the enquiry and alien to the intellectual realm of scientific history. It has practical significance in that, granted such a view of the matter, these moral judgments must be recognised to be an actual hindrance to enquiry and reconstruction; they are in fact the principal reason why investigation is so often brought to a premature halt. Yet we do not deny the importance of morality in life any more than we deny the hand of God in history, if we decide to conduct technical history without this postulate. On the contrary we shall find that, at the last stage of the argument, the historical realm emerges as a moral one in what we may regard as a higher sense of the word altogether.

Indeed we may say that precisely because all men are sinners and precisely because the rest of the truth about the matter cannot be disentangled short of the Judgment Day, the vindication of the moral element in history neither requires nor permits the separation of the sheep from the goats by the technical historian. Precisely because the issue is so important and precisely because life is a moral matter every inch of the way (while no historian can keep his ethical vigilance continuous or trouble to be making moral judgments absolutely all the time) – precisely for these reasons the occasional dip into moral judgments is utterly inadequate to the end it purports to serve. The effect of the whole situation is to make the judgments in question depend on the historian's unconscious selection of the moments at which he will think fit to raise the moral issue. There have been liberal historians who would tend to fix upon religious persecution as the topic which called for the particular act of judgment, as though it were the test case in ethics. It has even happened that the devout and reasonably virtuous person, who at certain periods would have needed exceptional saintliness or real originality of mind to break away from the prevalent principles of persecution, has suffered greatly at the hands of such historians, while the irreligious scoundrel, pursuing toleration from worldly-mindedness and motives of *realpolitik,* has been able to emerge with undeserved honours. Moral judgments are the more apt for this reason to be political ones in disguise – the historian is seeking to add a sort of supercharge to his condemnation of a certain policy. Some Whig historians would seem to have reserved the severest judgment for the men who support what they describe as 'absolutism'; and the rest of the wide world of moral action that is open to a man seems either to be ignored, as a mere matter of private life, or is reserved for a concession made in parenthesis. It is difficult to see how anybody who surveys historical literature – even leaving out of the reckoning the vast amount of writing which patently misuses the occasion for polemical purposes – could feel that justice is done to the place of morality in life by those spasmodic incursions into the field of ethics.

It would be foolish to take arms against the mere *obiter dicta* of historians, however, or against the incidental utterance of personal opinions. These things have a way of leaking into a narrative, and they are not to be taken as part of the structure of the history. We can accept them as the addition of a sort of colouring matter, while refusing to construe them seriously or to admit any pontifical claim that may be put forward on their behalf. A greater danger arises, however, if moral judgments are incorporated in the structure of the narrative, if they control the mounting of the story, and if they become embedded in the very fabric of our historical thinking. Above all it is necessary to resist those who claim for the historian the solemn role of moral arbiter, and particularly those who transfer this ethical preoccupation into the reconstruction of the whole course of ages. Important issues are raised if the struggle of Whigs and Tories is ranged into an epic conflict of the righteous against the unrighteous. Those who pitch their claim so high are tempting the technical student to a dangerous form of self-aggrandisement and do violence to the character of academic history. In their whole presentation of the human drama throughout the ages the conflict between good and evil is wrongly placed.

II

It may help us to keep our balance in our analysis of the question if we take note that even in the ordinary course of life the issues of morality are often settled with very little thought. Granting the all-importance of our ethical principles and ethical teaching – granting also the momentousness of the obligation that we are under to follow righteousness in our own conduct – we might wonder whether in reality there is any meaning or purpose in ethical judgments directed against actual people in respect of an action once that action has been done. In fact it is questionable whether any retrospective ethical judgment – I mean the kind of judgment which is directed against a personality – is worth anything, except in the form of the judgment that all men are, and men always have been, sinners. And this particular thesis owes its power and authenticity to the fact that in reality it is translatable for each of us into a self-judgment. For my own part I am willing to say that religious persecution is always wrong. It is wrong even when it is committed by people who are unaware that it is wrong. In other words I believe that it is essential to treat this planet as the kind of world in which men are allowed to 'choose' the God they will serve and the moral end for which they live. Granted that there is only one religion – namely the right one – this religion will be the one to suffer in the long run if that sphere of freedom is brought under the rule of force. I should hope, then, to go to the last limits along with anybody who refuses to countenance religious persecution for any reason. From this it will be clear that clarity or firmness in one's principles of conduct is not by any means the point which is at issue; and if there is question of a man who is about to persecute, we can all agree that he ought to be beset behind and before with the moral argument. From the same standpoint it is easy to say that an action is wrong therefore – the action itself is wrong even if the man who performs it is unaware that it is wrong. Such a judgment is almost a piece of tautology – it is merely a restatement of the original ethical principle. On the other hand, to denounce those who *have* persecuted; to condemn them in terms that suggest them to have been inhuman; to assume that they in turn did not have an ethical principle; and to assert or imply that they were more wicked than I am – all this is improper in itself and is a formidable obstruction to historical understanding.

As a preliminary to the discussion of the problem that concerns the historian, it will possibly serve a purpose to put forward certain general theses relating to the administration of moral judgments in the world at large. Such theses will help to define a mode of approach to this subject and will provide a framework for the argument that is to follow. If they give offence, however, they can be rendered otherwise harmless by the addition of the proviso that even if they fail to secure acceptance – even if a great wind comes to blow them all off the face of the earth – still, so far as I can see, this fact ought not to weaken the main argument which follows them, and to which they serve as a background.

The first point, then, is the belief that to some degree men are responsible for themselves and for their actions; but that all men are imperfect and that human suffering is greatly increased and multiplied by this general fact.

The second is the thesis that the difference between the wickedness and responsibility of one man and those of another, in the general world of nature (where it must be recognised that good fortune or adverse conditions play a great but still unmeasured part in the development of human beings), is so idle a question and so nice a point that it is not worth the wear of our fine intellects to discuss it in any imaginable conjuncture of life or history. Indeed, since human responsibility is so subtle a substance, presenting itself with vividness inside me, but not open to my vision at all inside another man – in other words, since I know that I could have done better than I did do while I can never tell what allowances I ought to make for other people – it is impossible to think one man essentially more wicked than another save as one might say: 'All men are sinners and I the chief of them.' It follows from this that moral judgments of actual people cannot defensibly or usefully exist in concrete cases save in the form of self-judgments.

Thirdly, though I, looking to the immediate future, must regard myself as a responsible person who may do things that are moral and immoral, and may follow or betray a law which is written on my conscience or a law that I have imposed upon myself; yet in regard to other people (who may think earnestly and differ from me about the law itself) and in regard to other people's actions once they are done (so that I cannot now prevent them), the passing of what purports to be a moral judgment – particularly a judgment which amounts to the assertion that they are worse men than I am – is not merely irrelevant, but actually immoral and harmful, not merely dangerous to my soul but unfitted for producing improvement in human nature anywhere.

Fourthly, granted that the State is under the necessity of punishing crimes, and granted that in the case of crime the offence is not merely technical but has moral implications (though sometimes the implications are not so assured or so direct as the world would like to believe), still we are not justified in expanding a legal verdict into a final judgment on a personality, or in assuming that because our own sins do not happen to have been also technical offences they are less morally blameworthy. If a man is sent to gaol, in fact, both the judge and the gaoler are to be interpreted as saying to him: 'Look here, old sport, we know that you may be a better man than we are, but since we can't tell what to do in order to save society, we have to resort to force.' If it is necessary to hang murderers, we must be sure that we are doing it because of a necessity and not out of moral indignation. And when we have done it we shall do well to reflect sadly on the bitterness of the necessity, and say: 'There, but for the grace of God, go I.'

Fifthly, since moral indignation corrupts the agent who possesses it and is not calculated to reform the man who is the object of it, the demand for it – in the politician and in the historian for example – is really a demand for an illegitimate form of power. The attachment to it is based on its efficacy as a tactical weapon, its ability to rouse irrational fervour and extraordinary malevolence against some enemy. As in such cases its efficacy is not lessened even when it is used unfairly and unscrupulously against those who have done no great harm, the argument for the use of this weapon is valid also for the unscrupulous use of it. The passage from the one to the other is indeed one of the most regular conjuring-tricks in the world.

Finally – so far as these statements of principle are concerned – I should say that, though I assume there are limits, I do not know where to place the limits to the operation of the truth that we condemn where we do not understand. This is tantamount to the assertion that the kind of ethical judgments which historians like Lord Acton have been so anxious to achieve are possible only to God.

All these principles are ultimately referable to the good old-fashioned doctrine that in the created world nothing really matters except live human beings. They would be inconsistent with anything like a Nazi view of human beings or the State or the *Volk*. They could not survive in a world that believed a society or corporation to represent the genuine whole, the authentic Person, with individual people as mere limbs of that body. They become submerged altogether if the State is regarded as the ethical end, and actual living people are construed as merely the means to an end. The principles imply that the biography of the worst of murderers could conceivably be written in such a way that he would be what in a technical sense we call the 'hero' of the story, so that our pity and sympathy would be around him as we followed him up from childhood. By the same argument what we should desire even in the case of a man like Hitler is not a mere angry denunciation. What would prove of incomparable value, if it were possible, would be an intimate account of him by a person who did not hate him too much – an account which would enable us to see how a lump of human nature (how a boy playing in a field) could ever have come to be *like that*. Best of all for the good of our souls would be an account such as would be written by a mother who was compelled to watch helplessly through the years while this child of hers took to terrible ways. But if it is objected that all this implies too delicate a regard for the effect of conditions on the development of human beings – if it is asserted that it seems to transform men into mere puppets, mere victims of circumstance – a counter-thesis must be stated in terms so uncompromising as to leave no doubt of a man's ultimate responsibility. It must be asserted that neither the Outcasts nor the pillars of respectability can in reality escape judgment if judgment is going to be pressed for. Indeed, if we are writing the life of a person whom we think the best of human beings, we may make him the 'hero' of the story, but this can only be in the technical sense again; we cannot pretend that he is a spotless saint. That is why we must go further than Lord Acton, who was inclined to feel that all the great men of history were bad men. If we want human responsibility we can only save it by something like the general dogma that levels all men – the doctrine that all are sinners, all are responsible for not being better than they are. In other words, none is completely excused if he has allowed even a bad education or the most adverse circumstances to corrupt his character. None is completely free and unconfined, but none is to be regarded as the absolute slave of conditions.

The principles that have been put forward would have to be defended in the last resort on the ground that these, and nothing less than these, enable us to do full justice to the authenticity of other people's personalities in a world where we cannot see inside other people. The theses would be inescapable if one went further and accepted the ultimate principle that no law of God or man, and no alleged utility, can supersede the law or transcend the utility of extending charity to all men, or can set imaginable limits to the law of charity.

Supposing, however, that these statements of principle fail to win acceptance, it may be pointed out that the issue with which we are concerned does not require that we shall commit ourselves to them at the moment. As a view of life they may be brushed aside, but the matter which does immediately concern us is the fact that in any case we must still adopt this point of view and transfer it into the very structure of our story of the past, the moment we undertake the work of the historian. In other words, if these theses and maxims are not true in the present world they are a necessary part of the structure of the realm of history. For in this latter realm, certainly, no ethical principle can be alleged to override the primary duty of extending charity, that is to say, increasing human understanding by an effort which always requires the expansion of sympathy.

III

In the conduct of life we shall often find it the case that the world condenses its assessment of a personality into something which it would not be unjust to describe as rather an aesthetic than a moral judgment. The question of a man's charm and his general demeanour, his bravery and the tact which he shows in the conduct of affairs – all these may be rolled up with our moral approval in what is really a judgment of the total product in so far as it is observable to an outsider. It cannot be denied that this over-all judgment of personalities – which is inclined to regard itself as a moral judgment, and as the final summing-up of the matter – tends to award medals to many of those qualities of personality which bloom more readily in the warmth of favourable circumstance. Here, where we applaud a certain nobility of mind or a niceness of disposition in a man, we are not unaware that fortune itself may have had a part in producing the attractive result. There, on the other hand, where we see forms like blasted oaks, we know that fine shapes of men have perhaps been devastated as by a flash of lightning. There is much of what we might call the broken crockery of human nature, where sometimes the fractures and deformities mark the effects of a not quite successful struggle with heredity and environment. Human beings are responsible at some point for the use which they make of the materials at their disposal – otherwise there is no sense in a discussion of moral judgments at all. But genuine ultimate assessments of worthiness are beyond the power of our mathematics to calculate. It is well that the ordinary verdicts of the world should be quasi-aesthetic ones and that we should realise them to be such.

Now the limitation of the historian, when he passes verdict on personalities, is that he is so liable to be satisfied in a similar way with the kind of judgment that is much less truly ethical than it pretends to be. Much of the benefit which is supposed to result from the whole practice is nullified by the deplorable fact that the moral judgments of historians are so often taken at a low level; we might even say that these things are almost invariably more rough-and-ready than anything else in the whole complicated fabric of historical writing. In reality they are pseudo-moral judgments, masquerading as moral ones – mixed and muddy affairs, part prejudice, part political animosity – with a dash of ethical flavouring wildly tossed into the concoction. They come blithely from impetuous adjudicators who have a rough idea of Henty heroes shining brightly against the background of something which is not cricket. Against such purveyors of rough justice Acton wrote much of his argument on this subject and made a justifiable protest; and it was in this context that he used the terminology of one who is protesting against a debasement of the currency.

In any case, in the world of pseudo-moral judgments there is generally a tendency on the one hand to avoid the higher regions of moral reflection and on the other hand to make moral issues out of what are not really moral issues at all. And if some may think that this is a laudable symptom of ethical zeal, they are mistaken. Even in the world at large it is rather a decline in genuine fervour which produces the desire to make gratuitous moral issues where such issues ought not really to be drawn. This is the method by which conflicts and controversies between human beings are rendered bitter and cut-throat and incapable of compromise. Against the argument that German historians in recent generations did great harm by avoiding moral judgments, it is possible to assert that these historians did harm, not through abstention and austerity, but precisely because they were not sufficiently reserved. They inserted pseudo-moral judgments and approvals, mixed ethics with reason-of-state, and tended to ratify and confirm Frederick the Great and Bismarck at the critical points. Indeed it has been commonly the case that those who were most eager to demand moral judgments in history – and who generally insist on the lower kind of pseudo moral judgments, since a more delicate handling of the matter would hardly serve their purpose – have on the whole been politicians, or men with a more or less direct political purpose, men with an axe to grind, who have rightly seen that moral judgments in history represent a considerable form of power. When Acton was confronted with the claim that the historian should forgo judgments on people, he betrayed on more than one occasion the fact that what preoccupied him was the fear that those bugbears of his, the wicked Ultramontanes, would in such a case get off scot-free.

Indeed, when the historian asserts the case for moral judgments on people, he is always bound to retreat and turn it into a case for the kind of verdict which pretends to be an 'approximation'. Only the latter – only the judgment which is mixed with a good deal of earth – will serve the militant purposes of the man who has a polemical intent. In any case, the lofty heights of the former are closed to the historian, whose apparatus and evidence are not qualified to carry him to such rarefied realms. The historian can never quite know men from the inside – never quite learn the last secret of the workings of inspiration in a poet or of piety in a devout religious leader. For the same reason he can never quite carry his enquiries to that innermost region where the final play of motive and the point of responsibility can be decided. The historian fails to pierce the most inward recesses and the essential parts of a man; and all he can depend on is a general feeling for human nature, based ultimately on self-analysis, but further enlarged in a general experience of life. Much can be achieved by a constant practice of that kind of imaginative sympathy which works on all types and varieties of men and acquires a certain feeling for personality. But the only understanding we ever reach in history is but a refinement, more or less subtle and sensitive, of the difficult – and sometimes deceptive – process of imagining oneself in another person's place.

Equipped with this general knowledge of human nature, the historian, when he deals with an historical character, examines external acts, thoughts that happen to have been put into writing, medical evidence, official records, perhaps the impressions of friends. Always he has to work on external data which he combines with all that he had previously learned concerning the interior of a human personality. He does not study human nature, therefore, in the way that an omniscient deity might observe it, with an eye that pierces our unspoken intentions, our thick folds of insincerity and the motives that we hardly avow to ourselves. It is true that an historian may feel that by imaginative sympathy he has almost completed the gaps in his picture of some historical personage, almost achieved what we might call an internal knowledge of the man. By great insight and by running all his molten experience into the mould that has been presented to him, he may feel that he has found the essential clue to a character – even to a man who has hitherto baffled the interpreters. Even this degree of knowledge fails, however, in that innermost region of all, which has to be reached before a personality can be assessed in a moral judgment. The genuine utility of such knowledge lies in the opposite direction altogether. In reality it adds more than anybody would have imagined to the wealth of historical explanation which can be assembled around a human action or a personality. And that enrichment of historical explanation is liable to be the thing which complicates or undermines the simple moral judgment which originally we might have been tempted to make.

When Lord Acton issued his justifiable protests against the slackness of the current ideas on the subject of moral judgments, the critical point in his argument had reference to the difficulty that arises from the inevitable incompleteness of the historian's knowledge. Rightly indignant against the casual handling of such important questions, he was wrongly severe, however, in the remedy which he proposed to administer. He approved the principle that we should beware of too much explaining lest we end by excusing the men of the past. Lack of knowledge in his opinion did not justify a suspension of judgment; he thought that the cause of morality would be better served if the historian erred rather by excessive severity. In the famous controversy with Creighton on the subject of the *History of the Papacy* it is apparent that the heart of the difficulty lay at the same point in the argument. Creighton was surely right when he said – after second thoughts on the subject – that he, for his part, could not bring himself to be the judge of Pope Alexander VI and must make allowances for time and circumstance. Acton, on the other hand, must have been right in believing that the historian does not know enough to exonerate such a man, and that, whatever might be discounted for the age of the Renaissance, the ethics of the New Testament had at any rate been in circulation for nearly fifteen hundred years. It is not to be presumed that either of these historians hesitated to agree that certain actions were wrong – that lying and poisoning were immoral. The dispute concerned the assessment of an historical personality, the verdict of posterity and of the last judgment – it was a question of the allowances to be made in the summing-up of a man. The truth was that Creighton could not know enough to exonerate. Neither, on the other hand, did Acton in reality know enough to condemn the man himself. It would have been better to recognise that the historian is not competent to make the necessary calculations, and that he carries the whole issue into a different realm of thought. It is not for him to steal the mantle of the ancient prophets; and it is more fitting that he should keep within the limits that his apparatus and evidence have set for him. Within these limits he has indeed a more important task; for Creighton and Acton had not even said the last word as yet towards the settlement of the question whether Alexander VI had actually committed all the deeds imputed to him. Beyond that, it is the real service of the historian to assemble all that can be gathered – everything relating to situation, circumstances, limiting conditions and preceding events – all that will throw a further explanatory light upon the conduct of this Pope. The historian will never be able to settle the controversy between Creighton and Acton or to determine whether Alexander VI had more than his fair share in man's universal sin – more responsibility than the rest of human nature. His function is to surround the man with all that can be gathered in the way of historical explanation.

Note

1. This chapter was originally published in *History and Human Relations* (London, 1951) pp. 101–30 (eds).

12
The Moral Framework of International Relations[1]

I do not know of any field which confronts one so peremptorily as that of international politics with the question: what is wrong with human nature? I personally join hands with those who believe that the all-important thing is to produce an improvement in human nature, through the actual deepening of personalities. One aspect of this – but certainly not the only one – would be the promotion of a profounder understanding, a greater disposition to try to understand human nature in other people, especially people not like-minded with oneself. At the same time, it should be made clear that the mere comprehension of the other party, or the other party's case, is not itself likely to solve the acuter problems of international relations. This particular aspect of the matter I should regard as the function of education or religion or moral discourse, though I would not deny that a statesman might contribute to it – might even inspire a fine ethical mood, a spirit of self-sacrifice, for example – in a nation – so that the moral factor can be important in the highest kind of leadership. But rulers and political leaders have to handle the world as it actually is, the world of what I might call ordinary human beings, and the statesman is not allowed to excuse himself on the ground that his policies would have succeeded if his predecessors had not spoiled the field, or if he had had a different problem to deal with or if his countrymen had been less egotistical. It has been a fault of modern liberalism, modern benevolent thought, that (much more than in its higher exponents in the 19th century) it leaves a gap in the argument – at the crucial point it ducks and says: 'human nature must rise to the demand; human beings must become better – or else!' In this context it means an appeal to a god from outside the machine; and the question how to capture that god is just the $64,000 question. The realm of international affairs is the one most calculated to suffer at one and the same time from the cupidity of the wicked and the unwisdom of the righteous. It is a field in which the problem of the self-righteousness of nations is more colossal and disastrous than the mere problem of national greed. I mention all this because I propose to have in mind the world of what I might call average human beings, without assuming that human nature in general is going to be changed overnight.

Starting from this basis, David Hume had a crucial instance with which he demonstrated the connection between moral conduct or ethical standards and the concrete arrangements of society. He pointed to the city-state of Genoa as it existed in his own day – a place notorious for the corruption of its political life; and he compared it with the contemporary Bank of Genoa, famous for its probity but run by the self-same human beings and involving the same stuff of human nature.

If morality can issue in this way out of the character or geometry of an organisation, it looks as though different planes or levels of existence can intersect, and perhaps one would as soon expect to hear poetry dribbling from the mouth of a mechanical concrete-mixer. But it seems that these are systems and institutions which, besides fulfilling their avowed functions in accordance with the blueprint, generate what I call the 'imponderables' – generate significant values in the field of human relations. I think that the intimacies of a small Cambridge college can have this effect; but so can the state, especially if, besides engendering loyalty, sentiment, habits and the consciousness of a common experience, it becomes a kind of moral order, so that people refrain from murder not merely because the law forbids it but because the prohibition has become an inhibition – so long as the system works it does not ordinarily occur to us to want to kill a man, or to take measures of self-defence on going out into the streets at night. It must have been in the early 1920s when a police-strike in Liverpool awakened considerable interest because it led to hooliganism and the looting of shops to a degree quite incommensurate with anything the police had had to deal with when they were actually on duty. Either there was a much larger criminal element in Liverpool than had been imagined, and the mere knowledge that there was a police-system had been sufficient to keep most of it under; or there were a great number of people who, under the routine of an ordered society, led fairly respectable lives without being conscious of any coercion, but who came off the rails when they discovered what men could do with impunity. The breakdown of such an order can make human nature so to speak worse than it really is (or can expose what it looks like in the eyes of God) by the mere fact that it opens up colossal temptations, it lands men too often in desperate corners, and it gives rise to fevers, hysterics and manias such as occur in times of revolutions and war. Perhaps when we consider even the European policy of Hitler we ought to ask ourselves how far we were responsible through creating conditions which offered such a lure and such opportunities – such as invitation to a political adventurer.

The question of the system (and of the imponderables to which the system gives rise) is important in the study of international affairs, if we are concerned not merely with disembodied principles but with moral ideas in their actual operativeness. The problem of the intersecting planes of existence becomes more difficult here, especially because sheer force has such obvious play – force which piles itself up until it looms mountainous, and which is so daunting because of a certain unanswerability that it possesses. History as we know it – the whole epic of the rise of civilisation – would be unthinkable if mankind had not been able to steal a march on the *régime* of force; but it is a case of stealing a march on it – not directly knocking it out – not merely confronting it with still greater force. And perhaps one has to use not only the virtues of human nature but also its weaknesses if one is to insert something of a moral order.

I suppose that, if one knew more about the subject, the strategies of the papacy in the middle ages might throw some light on the development of an international system or at least an international authority; for, in centuries of violence, when a bishop of Rome had no troops and the greatest potentates in Europe were hostile to his claims, he secured the destruction of his most powerful enemy and made a handful of monarchs (including the king of England) recognise that they held their dominions as papal fiefs. Also he secured universal sway for that system of canon laws which was to be so important a precursor of modern international law. I think he must have operated in part at least on the maxim that if you cannot do any better you must set a thief to catch a thief, set the wicked to catch the wicked and, the work being what it is, this is a maxim which might be given a limited – though only a limited – degree of commendation.

Certain points do emerge however from that European states-system which developed after the breakup of papal Christendom, and which endured in an attenuated sense down to 1914, providing the basis for what used to be called 'the old diplomacy'. It was shortly after 1700 that the French writer, Fénelon, in a thesis which virtually summed up the whole 17th century experience of diplomacy, provided a clue to a possible relationship between morality and force in international affairs. He showed that the moral conduct of states had a certain relationship to the distribution of power. He declared that if a state comes into a position in which it finds that it can do what it likes with impunity, it will in fact become an aggressor, even though it has been virtuous hitherto – even though it may have accumulated a lot of merit and earned the gratitude of smaller states by resisting a previous aggressor. If it happens to have a virtuous ruler at the moment when it comes to the top of the world, this may cause a little delay, but, says Fénelon, the good conduct will not last longer than a single reign. He clearly holds the view that the new situation will either bring the wrong kind of man to the top or produce pressures that a government cannot resist. Fénelon insisted therefore on a balance of power – insisted that the whole body of states (though each of them was a potential aggressor) should always combine, or should be prepared to shift their alliances, so as to prevent any power from attaining that absolute superiority. This was the worst predicament of all for everybody else, and its prevention therefore was the highest objective of diplomacy itself. Stop a Louis XIV, stop a Napoleon, stop the Hitler of 1940 – stop them before they get there. At a later date there emerged the point that if power was fairly evenly distributed – or if indeed a number of distinctly uneven states were kept in balance – governments themselves would tend to become more limited in their ambitions (too aware of the combinations that they might easily provoke against themselves); diplomacy and peace treaties would become more moderate, the general stability turning moderation into a habit, and not leaving such obvious openings, such big temptations, to political adventurers.

All this provided the skeleton-structure of what was the first theory of international politics to attain a general development through a succession of generations – appearing in publicistic literature, diplomatic papers, historical treatises, parliamentary debates, etc. Historians tried to trace the history of the states-system and even asked whether perhaps it was not unprecedented – the product of a unique set of conditions. It generated its own moral ideas, though these conformed to the power-pattern, to the basic geometry of the system; and indeed there was a good deal of moralising, though in a sense it could be argued that moralising was not necessary – the precepts were inherent in the logic of the system and governments just had to know this, just had to be wise enough to consider long-term interests as well as short-term ones. States should confine themselves to limited warfare, and, in particular, warfare only for limited ends. They should state the concrete thing they were fighting for – the precise piece of territory, for example – and not pretend that they were fighting to establish righteousness in the world. If you took to arms for a good reason – to defend yourself or to support the weak, you might well be fighting what was called a 'just war', but you must not turn it into a war for righteousness; and in any case the rights and wrongs of the struggle might change as the war proceeded – justice being measured not merely by the way in which the war began but by the question what kind of world would victory produce. Above all, you must never fight for the destruction of an enemy – that would create too much of an opening and an invitation to the next aggressor. In fact you, demanding more and more securities against the aggressor, may become the next danger to the continent before either you or anybody else is aware of it. You must always remember that the enemy of today may be needed as an ally tomorrow. In any case, do not let your war against revolutionary France become ideological, for that makes it a war of religion – not amenable to compromise. Anything like a war of religion is destructive of the states-system itself. Indeed, writers almost ran to poetry – they virtually came to singing hymns in praise of the states-system. It was as though all the states of Europe were members of the same club – you might jostle with one another in the competition for the amenities of the club – but your cupidity must stop short of the point where it destroys the club. It was the supreme claim that this – with its principle of the balance of power – was the only system that enabled small states to go on existing and to have a foreign world so open to the play of forces. It guaranteed even the internal *régimes* of more liberal states – England in particular pressing the further principle: non-interference in a country's internal affairs. And the alternative to all this was emphatically stated by the writers of the time – it was the Roman Empire, it was Napoleon, or, as we might say, it was a Hitler.

Now, when I look at the pattern of power at the present day – regarding this as the outermost framework of all – I find that the moral implications of this and particularly the question of generating the 'imponderables' are things not so easy to deal with. At the beginning of the 19th century the apostles of the European states-system were prepared to believe that this might ultimately become a global affair, but in their lordly way they probably understood this to mean that the rest of the world would gradually be drawn into membership of their club. There seems to have been a moment when, though I personally thought the European states might lose their predominance in the world. I still went on wondering whether they might retain as a residue of their former glory a kind of moral leadership freely accepted by Asia and Africa. As things stand I wonder whether there is not a pattern of power that pulls us in one direction, while an interesting development in moral ideas is pointing in a different direction. A lot seems to me to depend on future conjunctures, on questions of war and peace; and I wonder what would happen if, for example, one of the super-powers were rendered incapable of a strong foreign policy or a war by internal disruptions – giving another super-power clear headway if only for a month. Nor, in spite of the interesting developments in moral ideas, do I see any advance – perhaps indeed there is retrogression – in the field that touches just these high conjunctures, the attitude to the particular problems of war and peace. I might ask whether there exists today an international authority which achieves, shall we say in the realm of law, what the medieval papacy achieved. It is still more important to ask whether there is a global states-system sufficiently real to generate what I have called the 'imponderables'. I suppose one could conflate the two questions. Though UNO by merely bringing countries together and securing cooperative action, must be doing something to knit one with another, we may still ask whether, in its main operations, it gives the impression of making states and governments feel that they are members of the same club. And here the chief difficulty would arise if the sub-systems were more real to their members, more cogent in their influence, than the world-organisation as a whole – the clique of ex-Great Powers in Europe, the Afro-Asian group, the states behind the Iron Curtain – all of them developing their separate styles in diplomacy, their different evaluations and moral preferences. Alternatively, we might wonder whether, behind the amenities and the quarrelling there is not a conflict between two international systems, two concepts of an international order – an American one with something of the character of the older European states-system, and a Russian one, calculated to be more like Catholic Christendom in the Middle Ages. And certainly, if the effective division between nations is regarded by them as ideological, this introduces a loyalty higher than one's attachment to the states-system – something like a conflict of religion or a war between civilisations – it is fatal not only to the growth of the imponderables but to the notion of an international order. It may be useful to glance at something absolutely elemental, if only because it provides the only way of measuring how far the world has advanced in 3,500 years. If a few super-powers, Russia, the United States, and, say, China, for example, were confronting one another, each determined to shape the world as it wanted, and each attended by its satellite states, which were incapable of altering their régime or changing their foreign policy – then the overall predicament of mankind would not be analogous either to the Catholic system of the middle ages or to the European states-system of recent centuries. The real parallel would be the pre-classical empires (like the Hittite one), largely constituted by their patchwork of satellite kingdoms. So long as we have one party or all parties at the present day striving not only for peace but for a genuine *détente* – turning the idea of coexistence into a matter of principle, and insisting on the right of small states or peoples to choose their own *régimes* and decide their own foreign policy – this truly wonderful system stands as the final deposit, the real measure of the progress of the human race in international relations since the year 1500 BC – a progress chiefly due to the establishment and development of what I should call 'continuous diplomacy', leading to cooperative undertakings, and even the establishment of international organisations. I may be exaggerating the distance that the world has traveled. An Oxford scholar pointed out a few years ago the Hittites knew the doctrine of the self-determination of peoples. According to his translations, they had a picturesque designation for it – they called it 'the right of the cattle to choose their own stable'. The same writer even documents an occasion on which they applied the principle, though there are reasons for wondering whether they did it in a way that cost them anything. On the other hand, I do not think that I have found in modern history official documents that go so earnestly into moral issues as the Hittite ones – especially in regard to the origin of wars. I have never found another power that had such an aching desire to reason with its enemies, indeed with its fallen enemies, whom it so often forgave. Some scholars suggest that this was merely a cover for their own sense of weakness; but, even so it is an interesting way of making virtue of necessity – indeed, quite a creative use of diplomacy. In any case, there is independent evidence that the Hittites were pious and compassionate; and, for example, one historian after another has described how they were too squeamish to take hostages. For a lover of the Bible, the comparison with the barbarities of Old Testament warfare, centuries later, is the most stunning and sobering of all historical experiences. We might almost say that, in the realm of international affairs, precisely because it is so difficult to moralise force, the retrogressions are as astonishing as the advances.

Note

1. The manuscript original of this chapter is in BC File 110 (eds).

Part III
International Theory

13
Introduction

What follows are four papers prepared by Butterfield for discussion by the members of the British Committee on the Theory of International Politics (BCTIP). The Committee met fairly regularly between 1959 and 1984, initially at Cambridge University under Butterfield's chairmanship, and later at Oxford, Sussex, and London, with the primary purpose of discussing papers circulated in advance. It was funded by the Rockefeller Foundation, largely at the prompting of Kenneth W. Thompson who had been greatly impressed by aspects of Butterfield's work. Thompson already chaired a committee on international theory in the US funded from the same source, and he wanted Butterfield to attempt a similar enterprise in the UK. Excellent histories and interpretations of the BCTIP may be found elsewhere.[1] Suffice it to say here that the BCTIP's continuing significance resides primarily in the efforts of its members to explore the idea of an international system or society in both historical and theoretical terms serving, thus, as a precursor to what is now known as the English School.[2]

Despite being a driving force behind the setting up and continued operation of the Committee, and despite having presented at least nine papers to his colleagues, Butterfield's intellectual contribution to its work has been neglected. To be fair, those interested in his historical and religious thought have not been sure what to make of the Committee's work as a whole. Butterfield's interest in its work can be seen as one more distraction in a life of distractions from what, in various commentators' views, were Butterfield's big projects. International theorists too, however, have remained uneasy with Butterfield's attempts to make sense of international relations, in part from a religious standpoint in which, what might now be called, under-theorized entities like 'Providence' and 'imponderables' are assigned explanatory power.

There is some weight to all of these claims. In their different ways, the papers on morality and moralism in foreign policy above, and the diplomacy and balance of power papers below, stretch the meaning of theory beyond what the BCTIP (including Butterfield), probably, and the American side of the Rockefeller project, certainly, had in mind.[3] Nevertheless, in the four papers below delivered between July 1962 and July 1964, we see Butterfield addressing both the idea of international theory in general and central questions regarding international societies. The first, 'Notes for a Discussion of the Theory of International Politics', Butterfield presents as a stocktaking exercise. In it, he gives his view ('slogan') that political theory is 'the enemy' (because of its elevation of the state, rather than the individual) an airing, and asks to what extent a theory of international politics addresses the world as it is, as it might be if people's attitudes were transformed, or its own role in bringing about change. He concludes by calling for the Committee to re-focus on, and contribute to, the displacement of those very things that seem most familiar or are taken for granted in international relations. The paper is also of interest, however, for in it, Butterfield both holds up the 18th European states system as a model for how international order might be sustained and acknowledges that it leaves out much of what the Committee, or anyone else for that matter, who wishes to theorise about international politics must now consider.[4] To address this difficulty, he suggests seeing the world in terms of three 'theatres of action': 'the European states system, the modern international order; and the whole global scene'. While this seems to be an opening to the present interest in examining both the spatial boundaries between contemporary international societies and the temporal boundaries between contemporary societies and those they overlie (and may underlie), Butterfield moves off in another direction. The study of relations between societies will yield much of interest, but the internal relations of systems as closed off as the European states system was (if we can find them), will shed more light on the problems the BCTIP wishes to investigate.

This judgment is rooted in Butterfield's agreement with the historians of the Göttingen School and their Hanoverian precursors that the states system was a specifically European and rare and possibly freakish configuration of international relations made possible by the emergence of separate centers of state power within a common Christian culture with Roman antecedents. In his sense of its potentially aberrant quality, Butterfield shares much with contemporary leading figures in the English School, although there is little to suggest that he shares their sense that, potentially, the modern state system may be as fragile as it is rare.[5] He is much more interested in exploring its qualities than investigating its health and possible sources of it demise.

This is borne out in the second paper by Butterfield's focus on one of the institutions of the modern states system, 'The Great Powers'. The paper begins with a historian's concern for avoiding anachronisms. Great Powers, properly regarded as such, came into being with the emergence of modern Europe, Butterfield informs us. Hittites (although this is a poor example, given Butterfield's high opinion of them), other ancient empires and the Mongols, for example, could be regarded as great only in terms of size.

However, the nominalist and, indeed, discursive dimensions to Butterfield's thought rapidly become clear as he distinguishes between great powers and Great Powers. The latter presuppose a system of states mutually constituting themselves as a society of Great and Minor powers. Great Powers in general are conceded certain rights and burdened with certain responsibilities. A strong state may be regarded as a Great Power when it accepts both the idea of a system of responsibilities and that this system applies to it. Thus, not all strong states are Great Powers, while some that are no longer strong may continue to be regarded as Great, at least for some purposes. Nevertheless, Butterfield concedes that standing and cultural leadership are linked to political power. The relationship is complex. As Ranke noted, the Germany people were unable to exert their cultural leadership and weight until they were politically unified but, Butterfield comments, moral standing and cultural weight also add to the influence of the strong. Great Powers certainly can act oppressively but, on the whole, we are better off with them than without them and, indeed, the world in general would be a better place if men were prepared to make '... moderate concessions to moderate power'.

The system implied by the term Great Power has a social character in the sense that relations within it are regulated and interests moderated to some extent by a shared understanding on the part of its members about what is and ought to be going on. This theme is addressed in the third paper, 'The Historic "States-Systems"' in which the historic and plural character of such systems is introduced. States-systems (what we would now more likely call 'international societies') are defined by a shared understanding that, Butterfield argues, derives from a formerly existing political unity which its members are, in some sense, seeking to restore. They operate, therefore, at '... two levels of reality ... a universal system surviving only in its idea, and, on the ground, a realm of *Realpolitik* which seems utterly inconsistent with it ...'.

The origins and nature of universal systems are explored by Butterfield in the final (but earliest) of the four papers presented here, 'Comments on Hedley Bull's Paper on the Grotian Conception of International Relations'. Its point of departure is an argument made by Bull to the effect that many of the weakness of contemporary international thought and, thus, many of the ills of contemporary international relations flow from mistaken assumptions about the degree of international solidarity which exists in the world.[6] These mistakes result from positing a deeper moral code or natural law which is said to shape human conduct but which, Bull maintains elsewhere, simply does not exist.[7] In line with the positivist, or what he calls pluralist, tradition in international thought, Bull argues that there can be no more law or shared ideas about how to conduct international relations than that to which states are willing to accede for a mixture of their own reasons and motives. The implications are considerable for how we understand international relations and what we may reasonably expect of them. Should we, for example, be trying to restrict the legitimate occasions for warfare, possibly to the point of outlawing it, by punishing transgressors, or should we acknowledge that sovereigns will fight and seek merely to limit the collateral damage and suffering?

Butterfield begins by acknowledging that he shares many of Bull's concerns at the present state of international relations and thought about them, but rejects that both result from a concern for moral questions that is inappropriate to a consideration of the relations between states. The problems Bull identifies pre-date the rise of solidarist thought, according to Butterfield, even if they have been exacerbated by scientific and technical developments. And it is not moral concerns *per se* which have made things so much worse, but the sort of self righteousness that inspires crusades to create utopias in the here-and-now. In his reading of Grotius on just wars in which it is permissible for Christians to participate, Butterfield finds no grounds for associating him with such thought.[8] He does, however, employ Grotius to reject a claim he sees implied by Bull's argument, namely that reasons of states alone can generate an international order that is practically and morally acceptable. While we cannot ignore, or perhaps even avoid, the realities of power politics, Butterfield suggests, neither can we become complacent about them nor accept a code which sets the manufactured interests and priorities of artificial entities above the well-being of real, individual human personalities. 'I am prepared to start from the fact that international law, international theory, international morality, and the international order do not exist...' but, in a way which prefigures the emergence of constructivist thinking in international theory, Butterfield adds, '...this makes our discussion all the more important. We are creating these things, or refusing to create them, by the way in which we think about them now...'.

Notes

1. A summary of the workings of the British Committee by Adam Watson may be found at the English School home page, http://www.leeds.ac.uk/polis/ englishschool/watson98.doc. Files 30, 31, 159, 329, 330, 331, 332, 333, 334, 335, and 336 B.C. also provide the favour of how the Committee operated. A more analytical account is to be found in *Dunne*, pp. 89–135 and a historical account in Brunello Vigezzi, *The British Committee on the Theory of International Politics: 1954–85*, Milan 2005. Different dates are offered by different sources for the beginning and end of the Committee's working life.
2. For information on the English School see *Dunne* and the *English School* website http://www.leeds.ac.uk/polis/englishschool/watson98.doc.
3. Butterfield was not alone in this regard, however. His own preference that contemporary issues like 'the H bomb' be avoided by the Committee was thwarted by essays on subjects as varied as the consequences of the flight of Lin Piao, the press and foreign policy, and the attractions of Communism. International theory in its specifically North American form also receives Butterfield's attention in his notes on the 1967 Bellagio conference, BC File 109.

4. In his notes from the Bellagio conference three years later, Butterfield says '... wish to stress the fact that I hold no romantic ideas about the revival of the *ancien régime* or the patterns of eighteenth century Europe. I am really interested in the problem of method ... ' CUL File 109. As his lectures and other writings make clear, however, he cannot entirely avoid the contrary judgment. See, for example, his reference to the 18th Century as '... the Golden Age of Diplomacy ... ' in his modern European history lecture 'The Character of 18th Century Diplomacy', p. xviii, CUL File 328.
5. See, for example, Barry Buzan and Richard Little, *International Systems in World History: Remaking the Study of International Relations*, Oxford, 2000.
6. Butterfield's paper was a response to Hedley Bull's 'The Grotian Conception of International Society', a paper presented to the BCTIP in April 1962, BC, File 329. A later draft of this paper was subsequently published in Herbert Butterfield and Martin Wight (eds), *Diplomatic Investigations* (London, 1966), pp. 51–73. Bull also contributed 'The Importance of Grotius' to Bull, Benedict Kingsbury and Adam Roberts (eds), *Hugo Grotius and International Relations*, Oxford, 1990, 65–93.
7. For Bull's strongest expression of this view see his review of M. Donelan (eds), *The Reason of States* in *The Times Literary Supplement*, 28 April 1974.
8. It is perhaps worth noting here the way an exchange about how an earlier writer (Grotius) should be interpreted moves quickly from the historians' concern with making as accurate a sense as possible of the man in his own terms to the theorists' (or some theorists') interest in deploying him ironically in a debate centered on the present and the more recent past.

14
Notes for a Discussion on the Theory of International Politics[1]

I

I do not know whether it would be useful or might not in certain ways be disastrous, for us to reflect on what we have done, and, in particular, to ask ourselves whether we are getting anywhere – consider, above all, if we have any notion of what 'the theory of international politics' might be: I imagine that it need not do much harm and might draw our minds perhaps imperceptibly (or at least without conscious volition) into some useful direction, provided we determine to keep ourselves flexible, avoiding for the future any idea of a rigid project – hoping no more than that perhaps we shall have advanced to a greater degree of clarity if we return to a similar kind of stock-taking, say three years hence. We ought to guard ourselves even against any feeling that there is some urgent need for us to produce 'a theory of international politics'. It was my idea that some things (in diplomatic history, for example) might just be carried to a higher level of generalisation or thought out more thoroughly. On the other hand, our choice of subject might be affected, and our programme become perhaps slightly less a miscellany, if we now reminded ourselves of our original object.

II

Certain academic subjects have long been established: the study of diplomatic history, international organisation and international law, for example. I have often wondered (though without being sure that it was not vanity to start such wondering) whether we should consider 'the theory of international politics' as merely correlative to one of these – a sort of parallel or opposite number to international law – or we should imagine it as a system that embraces these and other things, at any rate holding a presidential position in relation to them.

If I take the only analogy I know, I think that the 18th-century theory of the states-system (which I regard as a coherent and fairly elaborate body of thought on the subject) concerned itself with the system as a whole, its working and its structure, as well as with the problem of policy. At the same time, it was close to the practice of diplomats and primarily envisaged the world of practice. In other words it was a theory of international relations, international political action, and would stand simply as a parallel to a system of international law. This effect (implying a certain self-completeness in the fairly closed intellectual system involved) may have been produced – and only made possible – by certain peculiarities in the theory itself.

We, as a Committee, however, have often found ourselves driven to ethical and philosophical discussions which would never have found any contact with the 18th-century theory of the states-system. We have stood on a fairly high platform and compared one doctrine of international law with another or tried to get behind the formal teaching of Grotius to his basic assumptions, his anterior outlook on the world. I take it that we should not wish too quickly to rule out the more comprehensive of the alternative ways of construing our subject.

III

Clutching at anything, in order to find an approach to the problem, I have wondered where we should be carried if we took it as our business to work out for the world at large a kind of international equivalent of that Political Theory which generally has reference simply to 'the State' as such.

Martin Wight, in the first of his papers, drew attention to the fact that there has been so little fine literature on the theory of international politics (or relations) to compare with the brilliant work that has been produced for centuries in the realm of Political Theory.[2] I remember telling the Committee, in the course of the discussion of that paper, how, for many years, I had made it my slogan that 'Political Theory is the enemy'. Political Theory takes hold of man's duties to his fellows or to mankind and comprehends them in his duty to Society or the State. It may not explicitly assert that there is only one Society or State, but it often argues as though only one existed. And the result is that thought tends to stop there. Even if something is added concerning the relations between states or between the citizens of various states, it seems to come as an unconvincing appendix, because thought has reached its apex in the highly-refined concept of 'the State'. In so far as this is true, I still have a sort of feeling that 'Political Theory is the enemy'. If I said that here is a traditional framework of thought that requires to be smashed, I should be in danger of taking refuge in mere exhortation. Also I should be raising the troublesome question: whether it is the object of our Committee to change the state of the world – whether we are conceiving a 'theory of international politics' which will only be valid when the world is transformed and men have had a change of heart. I would still ask, however, if – granting the world as it stands at present – one can have a theory of international politics, capable of driving the mind to any great height, without wrecking (by a blow at perhaps some small but crucial point) the framework of traditional Political Theory. In other words it would not be sufficient to make merely a few banal transpositions with the idea of adapting such Theory to a world conceived as international. This may mean (but I am not sure that it necessarily means) that a theory of international politics, capable of that fine blossoming which Martin Wight would like to see, is an impossibility, at any rate at present.

I do not know how far the philosopher, when he is discussing Natural Law, is affected by the fact that, on a given occasion, he is thinking of it in relation to the international scheme. I wonder, however, whether his envisaging a world of many states would affect his treatment of the nature, the grounds and the limits of political obligation, if that is a live issue. We seem to have found that a philosopher is important to us in many of the fields that we have opened for discussion. There may be a place on our Committee for a philosopher who is particularly appropriate for controversies that arise out of international law. Some of the problems raised by Hedley Bull[3] seem to me to be capable of carrying us very far: e.g. the question whether international law should apply to individuals or only to governments. But whereas I have an idea about the places where the philosopher has a special Role in ordinary discussions about 'the State', I am not sure whether it is possible to specify the key points at which the theory of international. politics particularly touches him.

By now, I am inclined to believe that we shall not be led very far if we consider the theory of international politics as analogous with any of the forms of Political Theory. Perhaps I am too accustomed to thinking that the fundamental problems are to be taken as already dealt with in the discussion of 'the State' as such. But in any case I don't seem able to find much connection between the actual problems of Political Theory at a certain level and those of the theory of international politics at a corresponding level. I always imagine that the study of International Organisations is the counterpart of what in our Historical Tripos we used to call 'Political Science B' (the study of the text and workings of the paper-constitutions of certain states), if it is not the counterpart of Constitutional History. Perhaps also it could just be said that we are bound to be greatly concerned with what might be called the 'Political Science' of international relations, or with the attempt to treat these relations in a more scientific manner. I imagine that if we chose to adopt the narrower rather than the more comprehensive view of our subject we should practically be confining ourselves to this. And included in this would be, e.g. the various elaborations of the doctrine of the balance of power, the whole question of the role of diplomacy or its special characteristics in one age and another (the effect of modern science on international politics, for example), the relations of foreign policy to military, economic and financial considerations, the technique of arriving at decisions, etc. Besides diplomacy, of course, there would be the problem of war.

Bacon makes the remark that there are some things which have become so familiar or which are accepted so automatically that people take them as self-evident, though they are just the things that are most in need of re-examination. He specifies among these the causes of gravity, circular motion in the heavenly bodies, heat, density, etc. He shows some insight in recognising that the progress of the natural sciences might depend on the pursuit of such enquiries; and I think that in fact the whole shape and structure of various branches of science were transformed by the sort of procedure he had in mind. One or two papers, like that on *Vital Interests*, brought Bacon's point into my mind, and I wonder how many other similar 'accepted ideas' (e.g. Hedley Bull's suggestion of *Alliances*, and possibly such things as *Aggression*) might not have the same effect: i.e. producing displacements over a greater area than the subject itself, and necessitating our getting other things square, so that we move towards an over-all intellectual system at this level.

IV

If we consider the theory of international politics at this level, the closest analogy that I know is the 18th-century theory of the European states-system; and, though I wrote something about this in a different connection, we did not discuss the matter, and I have always felt that such a discussion might be relevant to a purpose like the present one.

My notion of the European states-system (and the theory of it) springs partly from the (Gőttingen historians of over 150 years ago, and partly from the earlier publicists and practising diplomats from whom they themselves appear to have derived it. I believe they really opened the serious historiography of the subject, and, in fact, were the precursors of Ranke – so that much that has been criticised in Ranke was not really Prussian in origin, but rather Hanoverian, almost British–Hanoverian. In a sense they had the virtue of picking up the right end of the stick because they saw the European states-system as a phenomenon (even a freak-phenomenon) in global history, a thing so surprisingly unlike the colossal sprawling empires of the orient, for example. For them (as for the publicists and the practitioners at the climax of the 18th-century developments, and still more, perhaps, in the time of Napoleon) the real issue was between a Europe that should be unified like the Roman Empire and a Continent endowed with a map which was rather like a patch-work quilt. At the same time it was realised that the issue lay between a flat, uniform culture covering the whole area, and a system combining unity with diversity – culture as a dome of many-coloured glass. It seems to have been the view that the European system achieved unprecedented strength because power was consolidated at a number of regional centres; and that in fact, it was this system that brought Europe to the top of the world – a thing which called for historical explanation. Both the publicists and the historians seem to have been clearly convinced that the contemporary order could not have existed if there had not been the single culture and the more 'unitary' ideas of medieval Christendom behind it. In other words, a states-system could only come as a splintering of what had once been in a certain sense a unity. Heeren had the feeling, therefore, that if ever there had been an analogy to the European states-system, it must have been among the monarchies that succeeded the empire of Alexander the Great. Once a states-system existed, it was capable of extension, however, and it had actually been extended in recent centuries. Heeren in 1809 (that is to say, when Napoleon had virtually destroyed the system) seemed to have the idea that it might conceivably be capable of extension to the whole world. I suppose it is possible to imagine that if progress and expansion had taken place without cataclysm, something of this kind, gradually assimilating more areas into the system, would have produced the most favourable conditions for international development.

The very notion that Europe was a states-system, and not merely a congeries of states, had been greatly stimulated by the transition from the idea of the balance of power as something like a pair of scales to the conception of it as an equipoise analogous to that of the Newtonian solar system. It was this, even more than the element of common culture, which gave to the whole group of states the character of a system, possessing comprehensible internal interactions. The notion of the balance of power became therefore inseparable from the notion of a states-system, and the terms appear in fact to have been interchangeable – the balance of power was the structural aspect of the system. The force of the idea is not realised unless it is seen that it occupies in this 'science' the kind of place that gravitation had in the Newtonian system. The doctrine of the balance of power proved capable of considerable elaboration, and it seems to me that it came to stand as *the* 'theory of international politics' whether the 'theory' is regarded as an exposition of the existing order, or an explanation of actual practice, or an exhortation to envisage policy in a certain way. It was held that the states-system, with its balance of power, saved Europe from 'universal dominion', guaranteed the existence and independence of small states, and actually provided a significant Role for the foreign policy of small states. It was part of the theory that one should anticipate the aggressor – check a Louis XIV before he attained the position in which he would be able to commit offences over a considerable area with impunity. The policy implied making a great fuss over small acquisitions of territory that might tip the balance. As a result of this, both Frederick the Great and Burke have remarkable passages in which they declare the conditions of the time too unfavorable for effective 'aggression'.

It was realised that the system operated to secure liberty rather than peace. The whole theory involved the notion of warfare for only limited purposes;

but it did not take the prevention of war as its supreme object, and perhaps it tended to make some wars more terrible by collecting too many states into them, and securing too close a balance between the two sides. The chief 'snag' lay in the fact that there were elements making for rigidity in the international situation, so that states with traditional alliances might not be sufficiently agile when adjustments were required. But granted that the whole theory involved a process of abstraction – the kind of abstraction which is always necessary if one is to have any sort of 'science' – and granted therefore that there might be factors outside the system which would be of over-riding importance to us at the present day, I am not sure that we do not possess here an important 'model'. 'It seems to me remarkable to what a degree the maxims resulting from this system have retained their validity for any states – system, so that the breach of them in the twentieth century has had the consequences predictable, and sometimes predicted, in the 18th century.

I find that, as I think about the matter, it seems necessary to distinguish between three arenas or theatres of action., which may stand (at least in certain periods) one inside the other:

1. the European states-system
2. the modern international order
3. the whole global scene.

The European states-system, even as conceived by diplomats, publicists and historians in the eighteenth century, is a highly specialised thing, the result of processes of abstraction, and pretty well a closed and self-complete intellectual system, like Newton's explanation of the planets. Strictly speaking, the doctrine of the balance of power, in this particular system, should not even involve an element of exhortation such as might open a window to something else; it should involve no exhortation except that states should consider their long-term and not merely their short-term interests. But all this overlaps (in point of date) with what we call 'the modern international order', which we regard as emerging in the seventeenth and eighteenth centuries. And if I ask myself what is the difference between the two concepts or intellectual systems, I can only think that the latter contains all the things which the theory of the states-system eliminated from its picture – eliminated in the way that Galileo left such things as friction out of account when he discussed the path of a projectile. In fact the theory of the states-system is merely the specialised political aspect of an international order which had other aspects (moral, legal, etc.) and which, of course, might take up purposes other than that of liberty (freedom of action for its members). If a certain act is regarded not as impolitic but as illegal, if one entertains the idea of a general combination of states against an offender or if one seeks to erect an authority higher than that of the individual state, one is carried out of the beautiful automation of the states-system as men conceived it. But these two orders of things could co-exist, and either the political or the legal were just coordinate, or they were aspects of one and the same international order. If the object of the one system was to secure liberty, it would seem that the other gave the priority to peace; hence the stress on legality and illegality, on combining against an offender, on achieving a super state or a federal system or a League of Nations. And if the 'snag' in the theory of the states-system was its demand for greater mobility and flexibility than governments usually possess, the 'snag' in the other system is the tendency to rigidity. If it does not end in another Roman Empire, its legalism seems to result in an unfortunate freezing of the *status quo*.

V

In respect of the world outside historical Europe, I suppose that a good deal is known about the relations of the Roman Empire with the East in ancient times, and of Christianity and Islam in the Middle Ages. I remember reading the Spanish international lawyers in the sixteenth century on relations with the Indians, but, interesting though these are, I cannot recall getting any really new light from them on the problem of the relations between states and peoples. What would possibly be more important to us would be the internal relations of systems existing entirely outside our own – the question whether, for example, ancient Greece supplies anything more than mere analogies or whether the expert on China could open up completely new avenues for us. Joseph Needham[4] says that China had a conception of law so different from ours (which has tended to make the lawyer an unfortunate choice for a Foreign Minister) that its more flexible system might enable it to make a contribution – a happier form of international law than we are likely to produce.

Notes

1. A reference to Hedley Bull's paper on the 'Grotian Conception of International Relations' delivered in July 1962 (eds).
2. A paper entitled, 'Why is there no International Theory?' presented in January 1959 and published with slight modifications in: H. Butterfield and M. Wight eds *Diplomatic Investigations: Essays in the Theory of International Politics* (London, 1966) 17–34 eds).
3. A reference to Hedley Bull's paper on the 'Grotian Conception of International Relations' delivered in July 1962 (eds).
4. Joseph Needham (1900–95). William Dunn Professor of Biochemistry and editor of the impressive series *Science and Civilization in China* (1954–) (eds).

15
The Great Powers[1]

When we speak of 'Great Powers' we more or less consciously assume the existence of something like the internationa1 order of modern Europe. The meaning of the term has tended to become conventionalised therefore. What we have in mind is the society of nations which enjoy 'Great Power status' in such a system.

It is true that we sometimes tend to make facile analogies between past and present, as when we carry the term 'appeasement' back to the reign of James I. A number of historians have used the term 'civil servants' to describe a significant group of members of parliament in the reign of George III. Such transference ought to be made with one's tongue in one's cheek – with half a consciousness of the kind of anachronisms into which they are likely to trip the historian. And only in this limited sense – only without the use of the capital letter – are we permitted to describe the Hittites or ancient Egypt as 'great Powers'. Here the term is used as a mere description of size, and it does not in itself authorise the passage to any further line of reasoning. It would not, and it should not, normally occur to us to speak of the Mongol Empire or the Empire of Napoleon as one of the Great Powers.

It was said some years ago that Great Britain and the United States were seriously discussing whether they ought to go on dealing with France as though she were still a Great Power. It was decided that France should continue to have 'Great Power treatment', because, if this were withheld, she was likely to give a great deal of trouble. Supposing all this did not really take place, it is a thing that might have happened. And if the United States could discuss France in this manner, and could come to such a decision about the treatment to be accorded to her, then this country was possibly a Great Power after all, formidable if only by reason of the trouble which it was feared she might produce.

The term 'Great Powers', as it developed in the historical order of Western Europe, embraces things which in a certain sense may be described as imponderables yet are none the less, real, none the less effective even in *Realpolitik*. The concept should be regarded as peculiar to our traditional order – dependent perhaps not only on the existence or possibility of a balance of power, but also on there being not too wide a discrepancy between the greater states and the smaller ones. It would be conceivable that the United States and Russia, as what might be called 'Super-Powers' should transcend the older concept of 'Great Powers', producing a situation possibly more analogous to that which existed among the giants of remote antiquity. But, Great Power status being therefore rather like membership of a club, it is possible that these giants are still conforming to that status, in a way that the Mongol Empire and Napoleon Bonaparte never did.

It is precisely because of the imponderables which are involved that one may legitimately raise a fundamental question: the question of the 'leadership' of the Great Powers. Their leadership has possibly been more significant than their oppressiveness, so that we ought not to regard their existence as a kind of imposition on the rest of the human race. Because it is so natural to condemn the use, or resent the existence, of force, it is easy to be intolerant of the very idea of 'Great Powers'. One is tempted to reject the suggestion that anything in the way of leadership should be conceded to the strong. Our generation, in its hankering after the egalitarian principle, has tended to aspire to a democracy of states in which the authority of the units shall have no correlation with either their size or the stage of development that they have reached. It would please many people, and perhaps it would give satisfaction to something that is in all of us, if the weaker states of the world could prevail in all urgent and important matters by simply out-voting the greater ones in some kind of international organisation.

It may be unpleasant sometimes, and on occasion it may be dangerous, if the voice of Germany carries more weight in the world's affairs than the voice of Belgium – that of the United States more weight than that of Egypt. It is not clear that we should find the result more reasonable, however, if the will of the leading Powers in the world were simply overborne by the combined votes of the many states of Africa and Latin America. Historians have sometimes pointed out (or they have put us in a position to make the inference) that the leadership of the Great Powers may in fact be a matter of importance to the world. One example of this would be the part played by the leading states of Italy in pacifying that peninsula and keeping it in order in some of the later decades of the 18th century. Another example would be found if one examined the functions performed by the Great Powers in the period after 1815. In times and in areas in which such leadership is lacking, one tends to meet what the historian calls 'Balkanisation' in the realm of international polities. Apart from providing a kind of ballast for the whole system, the Great Powers, have improved human relations by imposing upon everybody at certain times a code of international conduct and etiquette. Irrespective of any charitable intention, and perhaps rather because of the nature of some of the vested interests they have had to defend, they have improved the intercourse between peoples and the quality of intercommunications. And they were moving towards leadership and a consciousness of responsibility when (even if primarily to serve their own ends) they began to tackle such problems as the Barbary Pirates, the internationalisation of rivers, the slave trade, the need for a Concert of Europe. The Great Powers are in fact a collectivity, and if, even so, they may become oppressive, a more sinister danger might emerge if they never existed, or if even they entirely exhausted their endeavours in their confrontations with one another. If we try to imagine their elimination, we have the vision of the possible 'Balkanisation' of the whole world.

In connection with all this, it is necessary that one should be on guard against the tendency to idealise small states. Once the Great Powers are in existence at all, it is inevitable that they should give the appearance of being the trouble-makers of the world. The competition at their level is of a crucial character, and on their struggles for existence or predominance the fate of many other things depends. It is tempting to infer that from their rivalries all the violence and the turmoil springs, that precisely from their great accumulations of power comes the scourge of war. The fact that all this is an optical illusion can be established by almost any examination of the conduct of the smaller, end even the smallest, states in European history. The truth is clear enough on those occasions when small countries are the clients of the greater ones. It becomes sometimes still more evident when the small states operate in an arena in which they are left to themselves. Small states certainly appear at times to be better than they really are, because, through the activity of greater neighbours, they may be in no position to behave in an aggressive manner. But if circumstances happen to provide them with the opportunity, they quickly show that potentially they are the same as the greater powers; the aggressiveness is there, though it may have been latent for a long period. Even in Switzerland in the early 16th century, the city of Berne began its aggressions and became oppressive to its neighbours almost as soon as it had emancipated itself from a local tyrant. It was part of the function of the greater states in fifteenth-century Italy to 'contain' the irresponsible rivalries and quarrels of the smaller states, whose conflicts menaced the larger interests of the country and jeopardised the general peace. In various fields the truth emerges that cupidity belongs to those who lose in the competition as well as to those who gain.

In the history of modern Europe, even the thing which we call 'power politics' has been something much more complicated than just a confrontation between magnitudes of blind force. In the emergence of Spain, France, Germany, Britain and Russia, it is not merely a military factor that has been important, or, to a certain degree at least, the military factor needs to be resolved into something else. Sheer force may be important, in that it gives a state a certain 'pull' in negotiation; and an unarmed country may resort to armament not because it wants a war but because it seeks to acquire this 'pull' in negotiation. But reason itself has a place in the transaction of this kind of business, and even today it would be possible for a small state to acquire influence by some novel initiative or some magnificent piece of persuasion. Harold Temperley, fresh from the Peace Conference of Paris, said that Venizelos had behaved with such wisdom that he fascinated the important statesmen and exerted a greater influence than the state of his country warranted. Though France appeared at the Conference of Vienna as the defeated country, Talleyrand secured significant successes there by virtue of mere diplomacy. After 1815 Metternich rose to the leading position in Europe, and he managed to exercise this leadership over a wider area and for a longer period than almost anybody else in modern history. It was not the result of any corresponding military ascendancy on the part of Austria, and in a sense it overrode what was felt to be the military predominance of Russia at the time. We can almost pinpoint the successive occasions when Metternich extended his leadership by feats of persuasion.

Great power status would seem to be accompanied both by responsibilities and by a sense of responsibility. The mere fact of being a large state is probably conducive to this; for where there are such extensive interest to be envisaged, there seems to be a constant need to consider the states-system as a whole, or the global situation. The small states – particularly perhaps the new nationalities in Europe and the undeveloped countries outside Europe – give the appearance of greater recklessness, and too exclusive a concern for local or short-term interests. Napoleon was in a certain sense reckless in foreign affairs; he was conscious that he worked for himself and not for either France or the future; he had a hunch that posterity would regard him as having been a scourge, like an eruption of blind forces in nature. But even in Richelieu, Frederick the Great and Bismarck one sees an awareness of the responsibilities of Great Powers and a sense of belonging to a society of such Powers. If the great states, through the extent of their vested interests, are often anxious to prevent even the outbreak of local hostilities which might lead to general war, they are also more concerned to calculate the consequences of decisions, since the burden of executing these is liable to fall upon them. It was once (and perhaps it still is) the custom to say – and I, forty years ago, was taught to say – that it is wrong to describe Prussia as a Great Power in the Revolutionary and Napoleonic period. She was more like a Balkan state, snatching profit wherever she could; and, owing to the genius of Frederick the Great, she had created the illusion of being a Great Power. In Britain, Austria and even Russia in the early 1800s, there is a sense of responsibility for the future in general, and for the whole order of things in Europe. When we say that a state has become a Great Power, we mean that it has joined a certain group and has qualified for a certain status in the international order; and it would seem that the recognition by others of that status is a significant factor in the case. The status even carries with it certain quasi-technical rights or privileges – the right to be consulted on general issues or to take part in certain kinds of discussion. The rights become rather difficult to separate from responsibilities, a point which becomes increasingly evident when the Great Powers establish a European settlement after a general war. It is possible that Russia, though she might be more dangerous, has today a greater sense of responsibility to the whole international order than either China or Egypt. By the Treaty of Westphalia, France had both the right and the responsibility of guaranteeing the terms which decided the internal constitution of Germany for a long future. When the Seven Years' War was over, Russia wished to establish her standing as a Power. She took the first opportunity to insist on being admitted as a guarantor of the Treaty of Westphalia.

In the conditions of historical Europe it has not been conceivable that a state should become a Great Power without being sufficiently advanced to be capable of cultural leadership. There is one feature in Ranke's thought to which it is not easy to reconcile oneself, namely, his emphasis on the idea that German culture and cultural influence became really effective only when the unification of the country provided them with a powerful political backing. If one remembers, however, the way in which the culture of the United States and of Soviet Russia were regarded down to something like the year 1945, it becomes hard to escape the conclusion that cultural leadership is not entirely unconnected with political power. The domination of French culture followed the rise of Louis XIV; and in the period before Louis, Spanish culture had had a remarkable influence in France. Taken as a group, the Great Powers after 1815 and even in the years before 1914, would seem to have operated as a whole to raise the standard of conduct in international affairs, and to advance the cause of civilisation, in spite of the crimes that they committed. If certain small states could devote themselves to cultural advancement without having to think about the game of power-politics, this in itself was a result of either the general or the local situation which the leading states had produced. The thing most injurious to civilised life might well be that universal 'Balkanisation' which would be likely to result from the absence of Great Powers. In any case, mere size would never have enabled the United States and Russia to become formidable – even to become nuclear Powers – if their culture had not been highly advanced. It is by no means clear that if the Great Powers abandoned the functions of leadership at the present day, the cause of civilisation would be advanced.

Richelieu says somewhere that small states have greater freedom of action than large ones and. that, when there are alliances between the two types, it is the larger that has the greater reason to fear desertion. Possibly there are factors in the present world – situation that give special leverage to the small state that is on the make, and so give further projection to the Richelieu thesis. The small states may be not only reckless but also calculatingly unscrupulous, relying on the certain knowledge that the greater ones will have to tidy up any mess they make. It would seem that the greatest opportunities for aggression and for violent changes in the territorial *status quo* now belong to smaller (or less developed) states which possess only the weapons of limited warfare. For the time being, they, rather than the greater states, can operate under the umbrella of that nuclear stalemate which, in a certain respect, paralyses the greater ones most of all. It would probably be felt that if the small states could acquire control of nuclear weapons, the situation would change considerably, because they would be more likely to be reckless in the use of them.

It is not easy to see how far there is real collaboration – and running collaboration – between the United States and its chief allies on the one hand, Russia and its chief allies on the other hand. It would seem that on the whole the giants of the present day have a great desire to be 'grouped', at any rate for general purposes. Apart from the fact that the principle is embodied in the Security Council, perhaps the best reason for saying that the society and the status of great powers still exists (in a way), is the fact that, because moral considerations are important, both the United States and Russia appear to want that system to exist. Both seem anxious to recruit the cooperation of other nations, and sometimes the competition between them is a race to secure the support of the uncommitted and undeveloped countries. This may tend to enhance still further the importance of the small states and the undeveloped countries, and sometimes it has seemed to lead America to policies calculated to undermine the position of the traditional Great Powers of Europe. All this is bound up with the paradox that if in a way naked power counts for more in the world than it did before 1914, moral factors also count for more in diplomacy than perhaps ever before. If, however, we have to see Russia and the United States as solitary 'Super-Powers' at one extreme and the small or undeveloped states at the other extreme, the Role of the traditional Great Powers (and perhaps some new associates of theirs) as an intermediate class might be very important, and the Role of Nasser in the Arab and African world, or of France in Europe may suggest that the members of this intermediate class need not be ineffective in their particular sector of the globe. If such as these may be 'Great Powers' (as distinct from the 'Super-Powers') we might say that it might be unfortunate if they abdicated their leadership at the present day.

In any case, the conclusion of the whole matter would be that much trouble might be saved in the international realm if men would make moderate concessions to moderate power.

Note

1. First delivered in July 1964. For the original manuscript see: BC File 330 (eds).

16
The Historic 'States-Systems'[1]

The Committee have been trying to discover how far anything analogous to the modern 'European states-system' has existed in other parts of' the world and other periods of history. Our studies have included ancient Greece, traditional China, and the great days of Islam and the special problem of medieval Europe. We have often examined the European states-system itself and we have touched on the Communist world as a kind of counter-system – a new form of international order setting itself up against our own. We may even have glanced on occasion at the possibility of a new states-system in the Arabian (perhaps even the entire Muslim world) or had the notion of a similar kind of thing (if not a bunch of separate regional systems) on the African continent. It is possible that (down to the present) however, we are lacking in information about two interesting if not crucial examples. One would like to know why Heeren – a man strongly impressed by the uniqueness of the European system – thought that the closest analogy might be found in the relations between the monarchies which followed the break-up of Alexander the Great's empire. On the other hand, if we are considering what might well be a special case – the system formed by a group of city-states – there is a chance that Sumer in the third millennium BC would reveal something fundamental, precisely because the development there is so early.

Our studies have sometimes confronted us with what were really tributary systems (or relations between some sort of super-monarch and his subordinates) and this may tend to confirm the old view of the uniqueness of the modern European order. But if it is these systems of subordination that are in question it is difficult to believe that any other documents will ever bring the topic down to its elemental features as effectively as Johannes Friedrich, *Staatsverträge des Hatti-Reiches*. For here, in the second millennium BC, the Hittite "Great King" sets out the *rationale* of the whole affair at great length, as though he were speaking to a child, and says, for example:

> II, 77. Those stipulations, however, are not based on receptivity but are imposed by the Hatti country.

and

> II, 135 If, however, you do anything evil, so will I, the Sun, do evil to you, and I, the Sun, will be free of every oath in the face of the gods.

The salient fact about the 'international systems' so far studied is that basically they do not seem to have been produced by the process of bringing together units which have hitherto been quite separate. The effective forces making for some sort of combination may be the elements of an antecedent common culture, but, though these may conceivably be strengthened by community of race, the startling fact is the importance of an earlier stage of political hegemony – a political hegemony which may even have been responsible for the spread of the common culture. Granted that a states-system is already in existence, it may not be difficult to add to it new units which were once outside it – even units that are of quite alien culture – as one might conceive of adding Turkey to the European system in the nineteenth century. But this itself does not seem to be very easy, and what emerges from our historical survey is an impression of the tremendous difficulty of actually creating an international order where no firm basis for it previously existed. It looks as though (in the conditions of the past at least) a states-system can only be achieved by a tremendous conscious effort of reassembly after a political hegemony has broken down. Perhaps this is why Heeren regarded the successor-states of Alexander's empire as the closest analogy to the European example.

City-states may seem to provide an exception to this rule, but it might be argued that they stand as a special case. If they possess, besides racial unity, a common language and culture (in whatever manner these may have been achieved) and if they exist together in a well-defined geographical area, they have surmounted some serious hurdles already. Fifteenth-century Italy possibly illustrates the fact that cities can move more quickly than nations to something that looks like a states-system. At the same time, it is almost impossible to avoid thinking of them as a 'system', even when we know that they have achieved very little of a positive movement towards such a thing. The surprising feature of the case is perhaps the fact that, in spite of their advantages, and in spite of their heritage of political wisdom, the ancient Greeks really achieved so little – far less than was necessary to save the order of things to which they belonged. In fact the city-state represents the stage anterior to political harmony (just as the states-system is the successor to it) – a point illustrated in ancient Sumer, ancient Greece and 15th-century Italy.

Perhaps because the various systems under examination belong to earlier periods of the world's history, the element of common culture which provides the unifying factor among groups of states would appear to be religion (or pseudo-religion). If the 'iron curtain' countries of today can be treated as a states-system, we should have to say that Communism functions in this combination as at least a pseudo-religion. It would appear that the cult or the creed supplies in this case not so much the supernatural sanction for the existing order as the cohesive principle itself – the very cement which holds the units together. In a similar way, religion is an important factor in the very birth and constitution of the nations themselves, and precisely because unity of religion brings cohesion; differences of religion may equally be responsible for separation – for making Croatia so different from Serbia, for example. The operation of this factor does not cease to be visible even in the modern centuries, as in the differentiation between Holland and Belgium and in the relations between England and either Scotland or Ireland. For the political purposes here in question it is not clear that Christianity (which involves certain potentially disruptive elements) is as serviceable as some other religions, or as a pseudo-religion, such as Communism. So far as a states-system is concerned it might be argued from the European example that, save for certain people like Grotius, religious fervour was an obstruction to development during much of the 17th century, while in the 18th century the cause was better served by the 'lapsed Christians', who, in this as in other respects would seem to have been an important factor in modern history. The secularisation of their ideals (with the admixture of rationalism) made it more easy for them to work for an international order as such, rather than for a Catholic or a Protestant order.

Polytheism had features which made it possibly important for the development of states-systems, at any rate in the period of city-states. In this connection, ancient Sumer is interesting since it seems the only case in which the pattern of the states-system was laid up in heaven, where the existence of hierarchical arrangements among the gods did not prevent their proceedings from having a parliamentary character. The real ruler in each city was the local god and the monarch here below was supposed to act as his administrator. Man was created to be at the service of the gods, and the products of economic life were brought to the temple, a proportion being granted to the worker for his own maintenance. If the human ruler of one city prevailed over the ruler of another city, he became the administrator on behalf of the god of this latter city, but he also looked after the interests of his own god, and each city would have subordinate temples for the gods of other cities. When Babylon, which had once been a relatively small place, achieved an unprecedented political hegemony, its god, Marduk, hitherto insignificant among the others, had to be brought forward to the first rank, and a new myth showed how, by his prowess, he had moved to the front rank among the gods. I suppose that, particularly with the dearth of evidence about the earlier history of the Greeks, it would be difficult for the modern mind to assess the importance of the alliances or fusions of cults in the creation of cities or the combinations between them. It would appear in general that in any case the bond (or the analogy) between the religious and the political orders was most complete in very primitive times, but that throughout very much of the whole long story religion has had a very profound effect. This may mean that, for the creation of a states-system (as in the emergence of cities and the birth of nations) the non-rational as well as the rational forces in human nature have to be mobilised.

Only to a smaller degree than might have been expected has the development of an inter-state system been stimulated by the need to combine against outside enemies. Even where the need was great and the development might have seemed most practicable, as in 15th-century Italy, this particular stimulus has proved insufficient. In spite of the conflict of over a thousand years between Europe and Asia, and the dreadful nature of' the incursions from the steppes, the Crusades depended on the existence of medieval conditions, and though the crusading idea may have given the pretext for plans of international combination, it does not seem to have been the effective factor in the story. The alliance between France and the Turks against the Habsburgs became a part of the balance that was achieved on the continent in early modern times. In the age of Grotius the religious motive and the desire to cope with the problem of warfare within Christendom would appear to have been the effective influence. Also, the most interesting developments seemed to occur after the Turkish threat had ceased. The powerful impulse seems to have been concern for the relations between members of the system—anxiety about the problem of war, but, above all, the fear of the powerful aggressor, the aspirant for 'universal dominion'. In a similar way, one sometimes feels that the fifteenth-century Italians were more afraid of the threat from Venetian aggression than of the barbarians from the north. Nothing helped Europeans to realise the existence of an inter-state 'system' so much as the idea of the balance of power, which is responsible for a good deal of the development that occurred. But even if the need to combine against a common enemy had been the decisive factor in the story, there must have existed already a certain sense of kinship (or a consciousness of inter-related interests) which recognised the common enemy as an outsider or an intruder.

When the large-scale systems are compared, one finds in traditional China and under the Caliphate as well as in medieval Europe the disposition to idealise the imperial order of things, and to go on thinking in terms of this order, even when the facts no longer correspond with it. The historian, therefore, is confronted by two levels of reality – in the air, a universal system surviving only in its idea, and, on the ground, a realm of *Realpolitik* which seems utterly inconsistent with it. One of the most interesting results of the attempt to use the comparative method on these cases has been the realisation that only in Europe did there emerge an acceptance of the facts as they existed at the *Realpolitik* level, and an attempt to build a new ideal on these, or to make a new kind of system on the basis of the material actually available. In other words, the existence of the separate nations came to be accepted, and their statehood recognised, while the idealised Empire came to be treated as the shadow that it really was. This acceptance of the facts as they actually existed seems to have been a necessary pre-condition for the improvement of the situation – for the conscious development of a states-system and an international order. While the notion of a universal empire was the ideal elsewhere, and Dante had been the spokesman of it in Christendom itself, the developments in Europe led before the end of the eighteenth century to the idealisation of the states-system as such, the valuing of an order of moderately-sized states, and even the prizing of small states for their own sakes – a disposition to treasure freedom and variety as ends in themselves, and to deprecate both unity and uniformity of the imperial ideal. Here is the actual inversion of the old dream or desire for universal empire.

The reason why in Europe men were able to face the realities of the situation and to adjust their theory to the facts was the thing which provided the main theme of debate after Dr Rosenthal's paper on the Islamic world. It appeared that in Islam the religion was more definitely embodied in the law and more definitely tied to a political system than in Christianity, which had long represented an opposition movement in the Roman Empire, had tended to emphasize the spiritual, and had made theology rather than law the 'acid test'. If, in spite if all this it had come close to caesaro-papism in Byzantium, the barbarian invasions and the downfall of the Roman Empire in the West had ensured the independence and even the predominance of the spiritual authority for a long time. For the modern international system as well as for the internal development of the separate states of Europe, the great conflict between Papacy and Empire would thus prove once again to have been a tremendous stimulus. These considerations may do less than justice to the resurrection of ancient political theory at the Renaissance and the application of the doctrine of the *polis* to the Italian city-states, as well as the similar influence of Roman law when it was applied in various countries or used in the separate states of Germany. It may not have been irrelevant that there were now alternative theories, close at hand, which could be given a more satisfactory adjustment to existing facts. Historians are generally faced with an aggregation of causes and conditioning circumstances, and rarely have the instruments for discovering which of these the effective one was, since so rarely can they tell what would have happened if a single one of the factors had been missing. The emergence of religious dissent that proved irrepressible must have been of momentous importance, since the principle of *cuius regio eius religio* meant that the *Respublica Christiana* was itself splintered and multiplied, and the secular power was recognised by treaty as having what was tantamount to unprecedented authority over religion. It is conceivable that if the Habsburgs and Catholicism had won a total victory in the 16th century, Europe since that time would have been saddled with fewer fantastic experiments in theology, and might have postponed much of the secularisation, besides avoiding many of the wars. But we might then have sunk into the stillness of an oriental despotism, clinging to ideals that often bore no relation with the facts – and this is perhaps why it is so difficult to imagine the religious conflict having such an ending.

Note

1. Delivered in January 1965. BC File 331.

17
Comments on Hedley Bull's Paper on the Grotian Conception of International Relations[1]

Although I am very much an amateur, and very much open to correction, I am not as yet convinced that Grotius is not suffering injustice at the hands of Professor Lauterpacht[2] and still more at the hands of Hedley Bull.[3] While feeling that I should probably agree with Hedley Bull in desiring a radical criticism of the conceptions that are now current, I am not clear that it is either true or useful to englobe these in the formula: 'the Grotian Conception of International Relations'. In fact, I still have to be satisfied that it would not be preferable to set Grotius up against the present-day views that Hedley Bull and I equally deplore. Though I dare not, on my present knowledge, cry 'Back to Grotius', I fail as yet to see why I should not take this line, especially if (as I suspect to be the case) this would be tantamount to calling Christians back to the better traditions of their past. If, however, I am wrong in all this, I merely duck and say that the 'Grotian' and the 'positivist' views are not the only alternatives. There is still a third intellectual system – the one that I *thought* Grotius represented.

The whole question may be made more difficult by the fact that there are inconsistencies and anomalies in Grotius, though I believe that a historian would find them more easy to reconcile, or at least to understand, than Lauterpacht does. In so far as they do exist, it becomes all the more important to allow for possible aberrations here and there in his work – i.e. not to assume that a particular criticism is necessarily fatal to the whole 'Grotian conception of international relations'. The fundamental fallacy is to equate the traditional doctrine of the 'Just War' with the modern doctrine which I would nickname the 'War for Righteousness'. For practical purposes at the moment we can say that the doctrine of the 'Just War' is the attempt to discover whether there is a kind of war in which a Christian can lawfully take part. And this seems to me to be Grotius's approach. Moreover he is pretty restrictive in his definition of what is permissible in law, and particularly what is permissible to a Christian. He allows for defence of one's own rights, and those of others; and he allows for the vindication of the whole system of rights (possibly saying more about punishment than I personally care for, though this is explicable historically, and not integral to a 'Grotian conception of international relation').[4] He seems to me to reject what I call ideological warfare, for he follows his predecessors in saying that it is not lawful to make war on other peoples because they are not Christian; and he is perhaps most significant of all in rejecting warfare on account of heresy. In the cases which he allows to be lawful, he insists in the fullest and most re-iterative manner that it may still not be the right thing to go to war.

He is remarkable in that he holds that an individual Christian should refuse to fight in a war that is unlawful (e.g. a war of aggression); but I see not the slightest tendency in him to confuse the 'Just War' with the 'War for Righteousness'.[5] Nor do I see any suggestion that the party which fights in a just cause is for that reason exempt from the rules. He is too clear about the issue from the start:

> Throughout the Christian world I observed a lack of restraint in relation to war, such as even barbarous races should be ashamed of; I observed that men rush to arms for slight causes, or no cause at all, and that when arms have once been taken up there is no longer any respect for law, divine or human; it is as if, in accordance with a general decree, frenzy had openly been let loose for the committing of all crimes.

It cannot be said that here his indignation about the causes of wars is more lively than his protest against their atrocities. And in the book itself he talks at tremendous length about procedures and methods – much less about the origin of wars. Far from taking the line that the origin is the only thing that matters (or that those who are right at the start will be excused for whatever they do afterwards) he is precisely the person (and, so far as I am aware, or can remember, he is the first person) to give a signal warning that the origin is not everything – that rights are created in the actual course of a war.[6] It is true that he says 'In war, things which are necessary to attain the end in view are permissible'; but this is a general statement which one is likely to find in any of the current views of international affairs, and Grotius went as far as anybody in enjoining the abandonment of the end itself, in preference to the continuance of the horrors of war. He also made a long list of atrocious things which could be done 'with impunity', listing the authorities for his view. But then, he says: 'I must retrace my steps, and must deprive those who wage war of nearly all the privileges which I seemed to grant, yet did not grant to them'. In reality he had been dealing with something more like positive law – things which he had to admit were in fact done with impunity or which he had to allow in law because 'coactive tribunals lend to them their authority'. Almost a whole volume is then spent in showing that these things are not just, or that there is a better way, or that Christians must behave differently. And surely

Lauterpacht (*Brit. Yearbook of International Law*, 1946, p. 39) holds this view of Grotius:

> In conformity with the view which has remained unchallenged and which is in accordance with the humanitarian character of his treatise, he lays down [Book III, Ch. IV, Para. IV] that the question of the justice or injustice of the war is irrelevant for the purpose of observing the rules of warfare as between the belligerents. Any other rule would add to the inherent evils of war the horrors of unrestrained licence and cruelty accentuated by what must often be an unverified *ex parte* claim to wage a just war.

At this point it seems to me that even the criticism directed against the twentieth century 'Grotian' may be unjust.

I particularly fail, therefore, to understand pp 5 and 6 of Hedley Bull's paper, though I think I generally agree with such sections of them as are criticisms of our own contemporary world. It seems to me that Grotius is with Oppenheim in regard to the three principles referred to, namely, those of 'necessity', of 'humanity' and of 'chivalry'. Again, I do not see why it is suggested that Grotius represents a fundamental departure from the basic Catholic tradition. The evils that are pointed out in this part of the paper seem to me to be the consequences of the modern doctrine of the 'War for Righteousness'. Even in regard to neutrals, I think that Grotius betrays a certain misunderstanding. Certainly he puts upon neutrals the duty of not assisting the party that has committed an unjust aggression. He allows them to join the party that had 'just' cause of war, but he does not by any means insist that they intervene in the struggle. In regard to the infraction of neutral territory, he allows this on the ground of extreme necessity, but does Hedley Bull mean us to understand that the positivists would differ from him on this matter? Indeed, it is Grotius who insists, among other things, on the point that this conduct is inadmissible unless there is 'not an imaginary but a real danger that the enemy will seize the place'. There is a sense in which Grotius says that if you have unjustly opened a war of aggression, nothing that you do in that war – not even the first soldier killed among the defenders – can be 'just'. But this is an understandable point – only objectionable when combined (as Grotius does not combine it) with the view that everything the other party does will be right.

All the same arguments apply to the statement on p. 7 that 'the Grotian doctrine of the just cause of war' involves 'all Powers in every war that breaks out'. It seems to me that Grotius is quite remarkable in the way he keeps insisting that even where the War would be 'lawful' or 'just' this does not mean that one ought to join in it. There is a place where he has been speaking of the duty of defending allies as though they were one's own subjects. He goes on to speak of the defence of friends 'to whom aid has not been promised, to be sure, but yet is owed under a certain principle of friendship'. He proceeds to say that war may be lawfully undertaken on behalf of anybody, because of 'the mutual tie of kinship among men'. Here he makes it clear that he means that a war is lawful if it is for the assistance of the weak. He then comes to a chapter which is headed: 'Nevertheless the obligation to undertake war may be disregarded without wrong'; and he begins by asking the question 'whether a man is bound to defend a man, or one people another, from wrong'. He quotes Sallust: 'All men who, when their affairs are prosperous, are besought to join an alliance for war, ought to reflect whether they may at that time maintain peace'. He wrenches a classical quotation to make it mean that one need not redress another's wrong unless it is to one's political convenience to do so [Book II, Chap. XXV, Para. VII]. Also he makes allowance at this point for the fact that even the victim may recognise the life of the aggressor to be more important to the world than his own. Certainly he objects to the conclusion of alliances when they involve a commitment to war irrespective of the justness of the war. I suppose he objects to offensive as distinct from defensive alliances, and I wonder whether there are fundamental objections to his view. I suspect that if correction is necessary here, it need not involve the overthrow of the whole 'Grotian conception of international relations'. Similarly, Grotius refuses to allow one to go to war with a state merely because that state *may* attack one later. Hedley Bull thinks that 'to defend ourselves effectively it is sometimes necessary to anticipate an attack'. Grotius writes on this topic [Book II, Chap. I, Para. XVII]:

> That the possibility of being attacked confers the right to attack is abhorrent to every principle of equity. Human life exists under such conditions that complete security is never guaranteed to us. For protection against uncertain fears we must rely on Divine Providence and on a wariness free from reproach, not on force.

But he adds [Book II, Chap. XXII, Para. V]:

> In order that a self-defence may be lawful it must be necessary; and it is not necessary unless we are certain, not only regarding the power of our neighbour, but also regarding his intention; the degree of certainty required is that which is accepted in morals.

I wonder if Hedley Bull's pregnant phrase about 'anticipating an attack' (which is in the dialect of the worst neo-Grotians) is intended to cover more than this. I suppose Grotius explains himself further in Book II, Chap. I, Para. V, when he says: 'The danger... must be immediate and imminent in point of time.'

II

In considering Grotius we must be careful not to allow Hedley Bull to bewitch us into forgetting that the atrocities of war (as the Introduction to *De Jure Belli ac Pacis* shows) are anterior to 'the Grotian conception of international relations' and not by any means its consequence. I often have the feeling that if I have been mistaken in my own thinking about the contemporary aspect of this problem, it has not been because I have been too 'Grotian' but because I may have underestimated the part played by sheer necessity in the intensification of the methods and means of war. I suppose it could be argued that when great masses of power are as equally balanced as in the First World War, victory could not have been achieved in shorter time or by less terrible methods, however wise men had been in their doctrine of war. In so far as this is true it may strengthen the argument that the proper way to deal with the problem of war is to tackle its causes, since there is so little hope of moderating its methods. At any rate I cannot at the moment, see any way of treating the problem of war, save by grappling with the causes of it, which at least are things that can be affected by changes in men's minds. For this purpose I think that I would attach the greatest importance of all to general changes in our thinking about the whole problem of international relations, and I imagine that here Hedley Bull and I are in agreement. But I can understand those people who feel, like Grotius, the appalling horrors of war, and who therefore hold that a terrible responsibility lies on the governments which, by what is called a 'political act', unloose anything like a great conflagration. I am not clear whether Hedley Bull wants to regard this 'political act' as outside the realm of morality or merely outside the realm of international law. He suggests that our ideas on international law should correspond with the times; but the condemnation of the initiators of wars seems to be the thing about which the contemporary world is most agreed. I gather that Oppenheim condemns 'a resort to war without pretext or reason of any kind', but I do not know the real bearings of this, or its relevance to any of the problems that worry me. I prefer *Realpolitik* to the 'War for Righteousness', but I do not see why I should prefer it to Grotianism or the Catholic tradition; and I want to know what it is that saves the 'positivists' from mere *Realpolitik*. I am prepared to start from the fact that international law, international theory, international morality, and the international order do not exist; but this makes our discussion all the more important. We are creating these things, or refusing to create them, by the way in which we think about them now.

I think my chief misgiving would be rather different from anything that Hedley Bull has suggested, though not dissimilar in one way, perhaps. I wonder sometimes whether it would not be wiser and more virtuous to recognise a moderate Role for force in international relations, on the ground that the refusal to do this seems to result in everything piling up for a colossal explosion. I do not know whether Hedley Bull (e.g. on p. 7) objects to the Grotian view because in the twentieth century it might tend to involve all powers in a war, all powers being supposed to intervene against the aggressor. The objection to the 20th-century view seems to me to be possibly the reverse one – the fact that in reality all powers will not join against the aggressor. On the other hand it is possibly more damaging to let things pile up into one colossal cataclysmic war than to have a continual series of small adjustment to changing forces and needs in the world. The objection to the 20th-century system is surely the freezing of the *status quo* if every resort to force is taboo. Whether inside a state or in the field of external relations, it often seems that some resort to force is needed to convince the world that the situation hurts or that one means to be taken seriously. Often I would say that some resort to force is necessary to have a matter placed on the agenda. In some cases a greater responsibility might certainly attach to those who extend a war than to those who first resorted to force. Even if we ever produce machinery for making reasonable changes in the *status quo*, I have a feeling that something of all this might still remain true. Perhaps one ought to bow to certain limited forms of the resort to force (which are often of a recognisable kind) and in some respects the effort to keep down the scale of a war might be more valid than the effort to abolish war. I can conceive of what I suppose would be limited forms of warfare, especially under the nuclear umbrella; and these might provide a necessary flexibility, avoiding the piling up of pressures which ultimately produce a colossal crack. Under such a system I suppose that Great Powers would have to be recognised as having a sphere of influence and it would be a case of not using the nuclear weapon against a power or a people which resorts to force to revise the *status quo* at the other end of the world. Somehow or other I think it may be desirable to recognise what might be a combination of force with reason in order to prevent what might otherwise mean a release of the entire world to blind force. But I do not feel sure that this need mean the 'overthrow' of the essentials of the Grotian system or that it would justify the positivists.

It seems to me that much of Hedley Bull's attack ought to be directed against a revolution in international thinking that has occurred since the rise of democracy. It is a reaction against both Grotius and the thought of the 18th century, and in a fundamental sense I suppose it is a product of 1789. It is the introduction of the idea of war as an instrument of policy in a sense more dangerous and insidious than was implied when mere *Realpolitik* was in question. Instead of being regarded as a reaction against an aggressor, or an attempt to right a local wrong, war comes to be taken as the appointed means for a general amelioration of the human condition, Hedley Bull says (p. 20) that 'to wage war for the international community is to wage war without limit'. If he means what I think he means, I disagree with him radically. The 18th century (which distinguished itself by putting the maintenance of the international order as the one thing superior to the pursuit or preservation of the interests of one's own state), is precisely the age which most developed the theory of limited warfare. It did this on the basis of the view that wars carried beyond a certain point, destroyed the international order itself. If you have an international order which avowedly comprises both Catholic and Protestant states – both monarchies and republics – war can be treated as a comparatively peripheral affair, for it is the conflict of religions and of regimes which is liable to be unlimited. In such circumstances, all the higher purposes of diplomacy (all the purposes higher than national self-interest) can properly be directed to the preservation of the international order itself. All this is changed if you have 'wars of peoples', wars that have become messianic, spreading democracy, whether other countries in general want it or not, and irrespective of whether they have reached the appropriate stage in their development. In one of its aspects this system of ideas is answered in advance by the Grotian thesis: 'An unjust cause of war also is the desire to rule others against their will on the pretext that it is for their good.' It might seem self-evident that the spread of Utopia is the best use that human beings can make of war; but there can be no doubt that it is the worst method in the world for those people whose purpose is to limit war. In fact, it has the effect of giving war a greater role than ever in history. The situation was bound to become much more serious when, in the Utopia that war was to produce, there was included the notion of perpetual peace, so that, besides the 'war for democracy', you had 'the war to end war'. And worst of all was the transition from the idea of the 'just war' – the idea of a war that may be 'lawful' but is never 'obligatory' – to the notion of the 'War for Righteousness', in which neutrality itself (as well as all idea of compromise) would come to be regarded as an offence. I cannot see this system of ideas in Grotius; nor do I believe that he would have accepted the decision to continue war (as a matter of avowed policy) to the point of 'unconditional surrender'. The logical conclusion of these comparatively recent developments in international thought – ideological diplomacy, the 'War for Righteousness', the battle for Utopia – is the policy of Soviet Russia since the last war. We need not go further than Grotius did in regard to heretics – that is to say in conceding that the other man's ideals, the other man's notion of the will of God, are as valid to him as ours are to us. The process surely compels us to recognise that if our recent popular ideas on diplomacy and war have been right, the Russians, the Communists, have been wiser and more admirable than the West. The trouble is that it is a reversion (behind all that Grotius and the 18th century achieved) to the large-scale 'war of religion'. It is tantamount to a war between international systems; and that is why it is more difficult than ever to have an international order at the present day.

Hedley Bull has called attention to two fundamental features of the work of Grotius – his assumption that an international system exists, and his basic individualism. Grotius still has a notion of Christendom, and says, for example [Book III, Chap. XII, Para. VIII]: 'Surely the bond which unites Christians is greater than that which united the Greeks of old, in whose wars a decree of the Amphyctions provided against the blotting out of a Greek city.' He lays great stress on the responsibility of the individual, insisting that a man shall refuse to serve in a war that he feels unjust. He will not allow that individual citizens can be punished for offences that have been committed by his King. He specifically rejects the kind of solidarity which burdens every single citizen of a state with a constructive guilt in respect of crimes committed by the Government. I am not sure that these things are inseparable from the general 'Grotian conception of international relations' which Hedley Bull has been criticising in his paper. But those who hold that the development of *étatisme* has been responsible for some of our evils may feel that these two aspects of Grotianism need revival rather than suppression.

It is true that one of the mistakes of the present day is the assumption that there exists an international order – with such claims on its individual members as an achieved international order might have – when such a thing is by no means an established fact. Grotius was able to think of a local system which corresponded with Christendom. The 18th century was conscious of the fact that its states-system was a regional affair – half-united before the sovereign states had really emerged, and vividly conscious of the Christendom or the Graeco-Roman culture that had cemented it together. While the ancient system was breaking up, new bonds were being created in international law and theory, as well as in the diplomatic operations of the European states, which could be conceived as knitting the sovereign states together. Early in the 19th century it was still possible to argue that the regional states-system of the European continent might ultimately extend over the whole globe. The mechanical construction of a League of Nations or UNO does not itself amount to an international order, and the evils which Hedley Bull has in mind are ascribable, in my view, to the fact that we have been assuming such an order to be actually achieved. But these international organs do have the effect of knitting the various member-states together in common action, and, for that reason, such bodies are worth while, in spite of the appalling inconveniences they might create in the meantime. Furthermore, I cannot accept the self-interest of a nation as the final objective for either an individual or a people. I still do not see why we should not say that over and above self-preservation and the promotion of national interests, the object of diplomacy should be the creation and the maintenance of an international order. This is at least better than saying that the higher object of diplomacy should be the promotion of Christianity or democracy or Graeco-Roman culture. In other words: If there is not an international order, it ought to be our object to make one, up to whatever degree the materials allow.

Finally, I do not know what ground we of the West can find to stand on (against the Communists) if we do not remain faithful to much of the 'individualism' of Grotius. I am not sure that I understand the basis or the implications of the charge that Grotius conceived of international law as operating between 'individuals' rather than between 'states'. He certainly thinks often of monarchs rather than states, and allows us to infer the rights and duties of Governments and states from illustrations that have reference to single people. I am prepared to believe that he may be open to criticism in this aspect of the question. On the other hand, when thought gets lost in abstract nouns and pretentious collectivities, I sometimes think that we might find our way back to sanity if we could reduce all questions to their simplest terms, doing our thinking on the basis that nothing save human souls really counts or really exists in this world.

Notes

1. Delivered in July 1962. BC File 330 (eds).
2. Cf H. Lauterpacht, 'The Grotian Tradition in International Law', *British Year Book of International Law*, 1946, 1–53 (eds).
3. Cf H. Bull, 'The Grotian Conception of International Society', in: *Diplomatic Investigations*, 51–73 (eds).
4. At the same time he is very anxious to let the tares stand rather than to pluck out any of the wheat. He insists that God lets great wicked communities endure for the sake of a very few good people living in them. He goes further than official Christianity has generally done in taking seriously the admonition 'Judge not that ye be not judged'. Also, he is very insistent that vengeance and punishment can be carried too far.
5. The chief exceptions in Grotius would be extreme kinds of barbarity and even in these cases Grotius adds the point: 'The wish to impose civilization upon uncivilized peoples is a pretext which may serve to conceal greed for what is another's.' In such cases of extreme barbarity, Grotius would even allow rebellion against a ruler and intervention by an outside power on behalf of the rebels. His position seems to me like that of a person who today would hold it as a general rule that powers should not intervene in one another's internal affairs but might think that Hitler's persecution of the Jews put Germany outside the international order, among the 'barbarians'.
6. Kelsey, translation, III, 600. 'A right is to be viewed as arising not only from the origin of the war but from causes which subsequently develop.'

Part IV
Diplomacy

18
Introduction

Butterfield was interested in exploring the possibilities for a scientific theory of international politics. Such a theory would clarify the constraints under which individual human beings exercised their free will and, in so doing, would provide corroborative evidence for Butterfield's emphasis on the virtues of sympathy and charity in seeking to understand the actions of others. However, engaging theory and, perhaps more importantly, engaging theorists also provided Butterfield with an opportunity to make his case for diplomacy and the importance of the assumptions on which good diplomacy and, hence, order in an international society rested.

Diplomacy was avowedly at the heart of the British Committee's efforts to develop international political theory.[1] Wight wrote of it as the 'master institution' of an international society.[2] Bull presented it as one of the principal institutions of the international society of states.[3] The Committee's principal published output appeared under the title *Diplomatic Investigations*. As Butterfield himself made clear in the BCTIP version of 'The New Diplomacy and the Historical Diplomacy', he and his colleagues were '... interested in the possibility of a "theory of diplomacy".'[4] The problem was being clear about what they meant by diplomacy. If it is what most professional diplomats spend most of their time doing, for example, then it is increasingly difficult to regard diplomacy as central to anything, especially to a theory of international politics. If, on the other hand, diplomacy serves merely as a synonym for international relations or was conflated with foreign policy, then the term either loses all its theoretical and analytical value or, worse, privileges as sovereign by sleight of hand the international politics of a particular epoch or a particular view of international politics.

The results, according to one commentator, are best characterised in terms of '... scholarly promise unfulfilled ...'.[5] In contrast to the stimulus provided to international thought by developing the idea of international societies as historical and conceptual phenomena, studies of diplomacy as a dialogical practice and attempts to explore the notion of diplomatic culture undertaken by people associated with the English School aroused little interest.[6] For the most part, they seemed to underscore diplomacy's dependent and secondary character, so much so that even the few people who tried to theorise about diplomacy felt themselves driven to examine it in terms of other things if they wanted to say anything important.[7] The exception among the Committee members was Butterfield. In the four papers which follow, he attempted to grasp the nettle involved in placing diplomacy at the center of international political theory, although he did so in a way which, until now, has elicited more criticism than approval.

All four involve discussions of the balance of power, its relationship to diplomacy, and the role of diplomats in developing and sustaining both. The first, 'The Development of Diplomacy', is a draft of a shorter piece published in a *Festschrift* for the diplomatic historian, David Bayne Horn.[8] While it antedates the other two pieces by several years, the theoretical argument is not as well-developed as it is a survey of the history of diplomacy together with a number of other international themes in which Butterfield was interested. It also, therefore, takes diplomacy's story further back than do the other two pieces that appear here. The value for our purposes, however, resides in three points early in the paper. The first is that, contrary to popular understanding, diplomacy, properly construed, is a historical phenomenon of recent provenance. Butterfield finds not a trace of diplomacy in the Amarna archive, for example, only the carping and bullying of great kings and their suzerains.[9] Where he does find precursors to diplomacy, in the Hittites' willingness to justify their actions and allow defeated enemies their own voice in peace agreements, for example, we find a clue to the second theme. Diplomacy not only has a moral significance, it also has a moral component. To act diplomatically is to act with concern for more than self-interest narrowly defined. The virtues implicit in diplomacy appear only episodically, however, until the establishment of resident embassies, initially in Italy and then the rest of Western Europe in the 16th and 17th centuries. Then they are elevated to systemic properties because a community emerges from the personal relationships created by continuous conversations among diplomats about shared problems. This brings us to the paper's third theme of interest here. The modern state system was in effect created, constituted and reproduced by the actions and reflections of diplomats – rather than states, governments or peoples.

One of the fruits of these reflections, according to Butterfield, is the realisation that while diplomacy always involves more than mere bullying and cajoling, it nearly always falls short of a reasoned discussion of problems of mutual concern. Power and interest always underlie what is going on. Diplomatic virtue, therefore, resides in seeking to substitute a virtual trial of strength and commitment by negotiation for a real test of arms, and working to pre-empt the achievement by anyone of a preponderance of power. Butterfield paraphrases Fénelon as declaring this to be the '... supreme purpose of diplomacy...' (amended to 'foreign policy' in the published version).[10] The paper then briefly reviews the rise and decline of the idea of the balance of power variously as an organising principle, a policy prescription and as a constraint upon policy, before noting the ways in which diplomacy can act as a power multiplier in foreign policy and peace talks. In the conclusion, Butterfield also praises Cold War diplomacy for dealing with unprecedented and difficult challenges. It has done so, in part, by rediscovering the wisdom of the balance of power relinquished between 1919 and 1939 when, in Butterfield's judgment, diplomacy reached its '... low water mark ...'.

The importance of the balance of power and what Butterfield judges to be the shortcomings of the men of 1919 when it comes to good diplomacy are salient themes of the other three papers presented here. Two are well known because versions of them appear in *Diplomatic Investigations*. On first reading, 'The New Diplomacy and Historical Diplomacy' sounds like an outburst against the rise of democratic and popular expectations in the 19th and 20th centuries. These expectations, in Butterfield's view, have contributed to the weakening of diplomacy and thus the international order and civilisation which it sustains, the whole sorry state of affairs culminating in the disaster of 1914 and the attempts of the men of 1919 to escape their history. He seems to be calling for a return to the verities of the 18th century, in part by educating people about the wisdom of the balance of power.

Certainly, Butterfield was a great admirer of the 18th century in this respect; in his modern European history lectures he would tell his students that it was '... the golden age of diplomacy ...'.[11] In this paper, however, there is a great deal more going on. It shows Butterfield wrestling with a number of questions. Is there a fundamental character to international politics existing irrespective of the nature of the regimes participating at any particular time? If there is such a character, does it impose laws on foreign policy and diplomacy which people can discover and ignore only at their peril? And if so, do these give rise to maxims about diplomatic behavior which can learned and taught and which people ought to follow if they are wise and good?

Butterfield looks back to 18th century Europe for one possible set of answers to these questions, but clearly, if reluctantly, rejects it on several grounds. 1914 decisively blew the world off the civilisational course set for it by 18th century Europe and, as he argues elsewhere, there can be no going back. Besides, the diplomacy of that period had its own shortcomings, engaging as it did in overly complex and subtle calculations which often lost sight of the proper objects of *raison d'état et de systéme*. It did so because, as Butterfield notes in an insight on which much contemporary English School theorising rests, 18th century international politics, not their 20th century counterparts, were the special case in the way they were insulated from other concerns. In contrast, 20th century international politics, even those of the nuclear age and the Cold War, are much more like business as usual in the way the 'moral factor' self-righteousness and passions are interwoven with them. Nevertheless, so long as '... policy is being operated within a system of nation states...' what constitutes wise and good diplomacy remains the same. Only our ability to recognise that this is so and the ease of following through on this recognition vary with circumstances.

Lurking behind the discussion of diplomacy, of course, and sometimes nudging its way to the front, is the idea of the balance of power, the subject of the third paper. In this, the tensions between Butterfield's scientific and humanistic conceptions of the dynamics of international politics become explicit. An international system may be described as a balance of power insofar as its members operate on the principle that no one, good-or-bad, should become so strong as to threaten the independence of the rest. On the one hand, however, Butterfield is at pains to present the balance of power as a historical phenomenon, the product of modern and European reflection on the problems posed by the relations of independent states. As such the balance of power emerges as a dependent variable, conditioned by contingent historical factors. This explains some of the chronic ambiguities and inconsistencies inherent in the concept. Earlier patterns of power, as Wight would call them, between neighbors merely remind us of the balance of power. They are not animated by the sense of middle to long-term self-interest that, Butterfield claims, allows states, or those who represent them, to develop a stake in the fate of the states-system as a whole and to moderate their ambitions accordingly. On the other hand, especially when he wants to emphasise the folly of ignoring the dictates of the balance of power, Butterfield uses the language of rules and laws insisting, as he did in the second paper, that the balance of power is not a prescription but '...rather a formulation of the way in which national interests actually do operate, and already operated before the principle came to its formulation...'.

What are we to make of this apparent contradiction in Butterfield's two views of the balance of power? One possibility is to make a judgment about where the weight of the evidence lies, but this is very difficult. Most of the time, Butterfield presents the balance of power primarily as set of ideas produced by sustained reflection at a particular time and place in history about a set of independently existing and relatively unchanging conditions. The conditions – the system of independent states – exist 'out there', but the response – thinking in balance of power terms, valuing the results, and acting accordingly – is rooted in the diplomats. Butterfield is uneasy with this, however, because it suggests that acting in accordance with the balance of power is merely a policy prescription which people are free to choose or reject. They are, but he wants to raise the costs of rejection by presenting the balance of power itself as something 'out there', if not existing independently then, once created, exerting a set of independent and punishing consequences if ignored. One can well understand why he wants to do this, but in the terms in which Butterfield was working, it cannot be coherently accomplished. In fact, it is not the consequences of ignoring the balance of power that must be feared, but of ignoring the problems which balance of power thinking was developed to address. As Butterfield notes elsewhere, the balance of power exists only to the extent that people believe in it, value it, and act in accordance with it. If it is not believed in then, the balance of power disappears. This again is illustrated in 'The Changing Moral Framework', the final paper which, despite its title, devotes considerable attention to the balance of power. In it Butterfield noted that this principle was difficult to apply after the First World War – a new situation having been created where 'those that were not victor states, had almost been removed from power politics'. Moreover, it came to be felt that prewar international arrangements – the conventions associated with the 18th and 19th centuries – had been too schematic and not sufficiently moralistic in their approach: that wars, henceforth, would be wars of moral indignation, fought to excess, not conflicts for limited ends. As a result diplomacy during the interwar period no longer served to modify or reformulate national objectives in line with international interests (i.e. stability): essentially, it became limited to devising ways in which the *fixed* objectives of states – whether territorial, political, or ideological – might be accommodated to some degree, generally without reference to the international system or the modalities of the former European order, of which the balance of power was such an integral and operative part.

Must we conclude, therefore, that a contradiction exists between the way Butterfield presents the origins of the balance of power and the way he presents its operation and consequences that is never resolved? Perhaps, but it is resolvable if we take an approach which was less open to Butterfield and his contemporaries than it is today. Rather than focusing on the logical consistency of his two presentations of the balance of power, it may be better to consider what he was trying to achieve. By the use of metaphors from a variety of worlds: the solar system; the dance floor; and everyday life, both Butterfield and his predecessors were trying to get people to think about international politics in terms of the balance of power. However, Butterfield was also trying to convey the weight and force with which a social system can operate even if it is 'only' the product of people's thinking in response to their circumstances. We may read the paper in Wendtian terms, therefore, as Butterfield's attempt to show what diplomats made of anarchy.[12] Their collective experience and reflection upon it led them to constitute the balance of power neither as a policy prescription nor as a mechanism by which a system of states might deal with certain problems, but as the international system itself. In this sense, instead of perceiving the balance of power as an autonomous element of foreign policy and international relations, Butterfield placed it within a more complex historical matrix, one that identifies distinctive lines of continuity between past and present power arrangements, their supporting concepts and articulations. Put differently, he sought to demonstrate the dynamic interaction, often tension, between the forces of tradition and newly emerging practices and ideas within the international realms, always remaining aware that, all too often, the effectiveness of diplomacy depends on elements transcending it, in part on domestic variables, in part interstate relationships. And of course, Butterfield maintains, if one values an orderly, but plural, world in which the liberty (if not necessarily the tranquility) of all, including the weak and small, is to be preserved, then the balance of power and actions consistent with it, remain your best bet. Let diplomats have their way and diplomacy can be happily conflated with foreign policy and international politics once again, giving rise to consequences which are beneficial for all.

Notes

1. *Diplomatic Investigations*, Preface and pp.11–12.
2. Martin Wight, *Power Politics* (Hedley Bull and Carsten Holbraad (eds), (Harmondsworth, 1979) p.113.
3. Hedley Bull,'Diplomacy and International Order' in *The Anarchical Society*, (Basingstoke, 1977).
4. 'The New Diplomacy and the Historical Diplomacy', BC File 329, 23-26 September. The first paragraph with its reference to 'a theory of diplomacy' is missing from the version published in *Diplomatic Investigations*.
5. Iver Neumann, 'The English School and Diplomacy: Scholarly Promise Unfulfilled', *International Relations*, 7, 3 (2003), pp. 341–69. cf I. Hall, 'Still the English Patient? Closures and Inventions in the English School', *International Affairs*, 77:4 (2001) pp. 931–42.
6. See, for example, James Der Derian, 'Hedley Bull and the Idea of a Diplomatic Culture' in Rick Fawn and Jeremy Larkin (eds), *International Society after the Cold War* (Basingstoke, 1996). A revival of interest in the idea of diplomatic culture is underway, however. See Geoffrey Wiseman, 'Pax Americana, Bumping into Diplomatic Culture', *International Studies Perspectives*, 6, 4 (2005), Iver Neumann, 'To Be a Diplomat' in *International Studies Perspectives*, 6, 1 (2005) and Paul Sharp, 'The Idea of Diplomatic Culture and Its Sources' Slavik H. *et al.*, *Organizational and Professional Aspects of Diplomatic Culture*, Diplo Projects (2005) and 'Herbert Butterfield, the English School, and the Civilizing Virtues of Diplomacy', *International Affairs*, 79, 4 (July 2003).
7. See, for example, Christer Jönsson and Martin Hall, *Essence of Diplomacy* (Basingstoke, 2005).
8. Ragnhild Hatton and M. S. Anderson, *Studies in Diplomatic History: essays in Honour of David Bayne Horn* (London, 1970).
9. The Amarna tablets were discovered in the 1880s. They record correspondence between the pharaohs of the Amarna period and their neighbours. Butterfield's reservations notwithstanding, they are regarded as one of the earliest diplomatic archives. See Raymond Cohen and Raymond Westbrook (eds), *Amarna Diplomacy: the Beginnings of International Relations* (Baltimore, 2000).
10. François Fénelon (1651–1715), credited with the earliest explicit account of the dynamics of the balance of power between states in his *Instructions*.
11. 'The Character of 18th Century Diplomacy', p. xvii, BC File 328. However, in his notes from the Bellagio conference Butterfield says'... I wish to stress the fact that I hold no romantic ideas about the revival of the *ancien régime* or the patterns

of 18th century Europe. I am really interested in the problem of method', File 109 BC, 1967.

12. Alexander Wendt is widely credited with bring the notion that social reality is constructed to international theory. See, for example, *Social Theory of International Politics* (Cambridge, 1999) and 'Anarchy is What States Make of It: the Social Construction of Power Politics', *International Organization*, 46, 2, Spring 1992.

19
The Development of Diplomacy[1]

I should like to begin by just referring to two examples from remote antiquity – both of them rather a puzzle to historians.

The first is the case of the Tel el-Amarna tablets – 350 letters connected with Egypt in the period going back as far as 1400 BC; the most renowned diplomatic papers of pre-classical times. I have poured over them at dead of night and can't find any sign of diplomatic craft in any of them. The period was one when an enormous importance was attached to the gifts that passed from one friendly ruler to another. Many of the monarchs seemed dissatisfied with the gifts they received – but all they seemed able to do was resort to bleating and begging, or they imprisoned the messengers who brought the offending articles. There was one doubtful passage, perhaps. A ruler was supposed to receive a solid gold statue from the Pharaoh of Egypt; but when it arrived it was lightweight, so he had it examined; and it turned that it was made of wood and merely plated with gold. He told the Pharaoh that he ought to be more careful in future about superintending the conduct of his officials. I wondered for a moment if perhaps he wasn't being diplomatic in so delicately putting the blame on the official rather than on Pharaoh himself. The official clearly was the culprit, however.

But earlier than this the Hittites had been building up an empire from their center in Asia Minor – and they have left foreign policy material which has provoked a good deal of discussion. They seem to have thought it necessary to provide historical – even ethical – justification for the wars they undertook, and they did it without insulting their enemies in the way that nations of the 20th century have done sometimes. They did it in lengthy preambles to their peace-treaties and scholars say (though I haven't personally seen an example of this) that they allowed the defeated enemy to draw up his own account of the origin of the war for his own copy of the treaty – conduct which one or two people have contrasted with that of the allies imposing the War-Guilt clause on Germany in the treaty of 1919. When a vassal king had rebelled and been defeated, they liked to forgive him but they treated him to a lot of moralising, urging him to remember what he owed them and

I've never seen any other diplomatic instruments that so harked on the duty of showing gratitude. There is now quite a literature which sets out to show that the famous Covenant between God and the children of Israel follows the pattern of those treaties between the Hittite Emperors and their vassal kings. One or two scholars had suggested that the Hittites in spite of their victories might really have been weak – disguising their sense of weakness with a show of moderation; but if this really was the case one has to give them credit for their diplomatic activity – for the way they made virtue of necessity. I know of no other case in which documents of foreign policy so regularly bring a moral tone into international discussion – great emperors apparently moved by such an urgent, such an aching desire to reason with their enemies. But there is too much evidence of the actual mildness of the Hittites – quite a contrast with the Hebrews of the Old Testament hundreds of years later – they really didn't like killing whether at home or abroad. They were too squeamish to kill hostages. An Oxford scholar has shown that they knew the principle of the self-determination of peoples – they called it the 'right of the cattle' to choose their own stable. I think one would have to say that the Hittites had their own style in diplomacy.[2]

For a very long time, diplomacy was not conducted by ambassadors regularly sent to reside in a foreign country. For special occasions some magnate or court dignitary or officer of state – even perhaps a high ambassador – would be sent out on some particular mission: to end a war or settle a quarrel or make an alliance, even to attend a coronation abroad. An important part of this task would be to deliver a set speech at the court he was visiting – rather in the way that our foreign secretary or the American representative gives a speech at the United Nations. We have the text of such speeches given in the 16th century but also we have thumbnail accounts of such things which go back for thousands of years. There is a picturesque example in the Old Testament, which would be illuminating even if it were fictitious, because it throws light on an important aspect of diplomacy. It belongs to about 701 BC when the kingdom of Judah was a tributary state, subject to the vast empire of the Assyrians; but it had revolted against that empire, having relied on assurances of help from Egypt. The Assyrian emperor Sennacherib first disposed of the Egyptian army and overran the kingdom of Judah up to Jerusalem itself, which was now expected to be besieged. However, he sent three emissaries to the king, to induce him to surrender without a siege; and these men met three delegates of the king of Judah, King Hezekiah, the encounter taking place in the open air. The Assyrian spokesman first taunted King Hezekiah for going to war when he could do nothing without the horsemen and chariots of Egypt. He pointed out that no help could come from the Egyptians now and asked the Hebrews if they thought that 'meek words' could win a war. He then said that his master would offer them a wager. 'I will give you 2,000 horses if you for your part have even got 2,000 men who could ride them.' He added that it would be useless for the Hebrews to hope for help from their own national god. He said, 'It is your god who told us to come.'

They were in the open air, and some of the Jewish people were on the walls not far away, listening to what was going on. The leader of the Hebrew delegation asked the Assyrians if they would speak in Aramaic – then a sort of international language in that part of the world. He said: 'When you speak in Hebrew these people on the walls can understand you, and that is inconvenient to us.' Whereupon the Assyrian spokesman, instead of obliging him turned to the onlookers and made it clear that it was they whom he wanted to address – he wanted to talk to the populace rather than the government. He told them that their king was treating them unfairly; they would be the ones to suffer if he insisted on continuing the war. It would be far better for them if Jerusalem just surrendered; for then they would at least save their lives, and for the time being they would be able to eat of their own vine and fig tree. He admitted that they would be carried into exile later – that was the policy of the Assyrians when they conquered a country. But, he said, in his coaxing manner you will be taken to, 'a land very much like your own land, a land of grain and wine, a land of bread and vineyards, a land of olive trees and honey'. And, he said this would at least be better for them than death. In any case, it was no use their relying on their God. Look at all the nations that trusted in their gods but still were beaten; 'where are their gods now?'

Such a mixture of reasonable argument with hoaxing and cajoling – and the additions of taunt and threat – reminds one of Napoleon when he was in a similar position and something of the same can be found in Hitler.

The narrative establishes the first point which has to be kept in mind by the student of international relations. If one begins by thinking of diplomacy as an attempt to solve problems by mere reasoning, mere discussion – all on the assumption that the better argument will prevail – one is liable to be led seriously astray. For a beginning, at least, it is better to pick up the opposite end of the stick, squarely confronting the fact that something more than the mere difference of opinion is involved, something more than the mere case of understanding, or failing to understand, the other parties' point of view. Initially at least, the need for diplomacy arises because there is a conflict of wills or collision of interests, each party being anxious to get its own way, even though it knows all the arguments and *does* understand the other parties' point of view. It is a case of operating on men's wills and even on their willfulness – getting them to agree to do something which they have previously determined not to do. You have to introduce a more dynamic element into the transaction: to discover what will be the effective inducements. Negotiations work by means of promises, by the offer of a *quid pro quo*, by any kind of appeal of self-interest – attempts at arranging a bargain by means of cajolery. But also they have resorted sometimes to taunts and bullying, to quiet blackmail, to bribery, to impudent bluff. They may even threaten war and that very threat may itself be a piece of bluff, though you on the other side may not be sure of this. And, of course, a giant like Russia, will generally have greater bargaining power than a state like Belgium. Here is a situation where trade unionist leaders sometimes understand rather better than other people. They know that if they were deprived of the ability to strike or to threaten a strike – if they were reduced to having to convince employers by sweet reasonableness alone – they would be in a pretty poor position for negotiating. In earlier days diplomacy ran very much to sensationalism and melodrama – there were too many Machiavellian ruses and confidence tricks. You might perhaps say that diplomacy is better than having an actual war.

And of course it did become more refined, more civilised in the course of time – though without losing its initial, central nature. The greatest factor in the development that occurred was the rise in the 15th century of the practice of sending out a regular ambassador to reside in a foreign capital – a man who would stay two or three years in one country and then be succeeded by someone else. There is some evidence that the purpose was to obtain a more continuous supply of information, and learn what the foreign country might be up to; and certainly the courts that disliked receiving ambassadors were governed by the fear that it would be like taking a spy into the bosom of one's family. Of course the magnates, court dignitaries, ministers of state and high ecclesiastics were not to be so easily available for prolonged periods of residences abroad, and a rather different kind of official emerged – a man who would be ambassador in a capital for two or three years, but then be moved to another and make a career of it. Books came to be written about the profession, manuals and text books for the guidance of ambassadors. For a few years at the end of Louis XIV's reign they established a special college in France for the training of diplomats.[3]

But at first it was thought to be a poor sort of job. You went to live in a foreign capital and sometimes you were suspected as a spy, you might even be surrounded by spies yourself. Occasionally the country in which you lived – especially in the public service – would be forbidden to have relations with you at all. If anything went wrong you were liable to be badly supported by your own government, and in any case you were badly paid, you had insufficient office staff, and you were likely to be out of pocket at the finish. There are cases of legalities being passed – in Venice for example – to penalise a man if he refused to go out as an ambassador after being asked by his government to serve. And then, when there was an important piece of business to transact, the ambassador might be superceded by a high officer of state sent out on a special commission – rather in the way that our Foreign Secretary fly out to Moscow today. Some ambassadors found it boring to live in a foreign capital – I have an extract from a letter from one who even found Paris tedious. Also diplomacy itself could still be a somewhat disreputable affair. In the reign of James I, an English ambassador, Sir Thomas Wotton, coined the definition that an ambassador is a person who 'lies abroad' for the good of his country. So long as he expressed it in English, he was covered for he might have been making a pun – 'lying abroad'. But there is a report that he wrote it down in Augsburg and lost a job, because he translated it into Latin – *ad mentiendum* – and that could only mean telling untruths.

Yet the introduction of these resident ambassadors did bring about the biggest change that ever occurred in the conduct of international politics – though it was a change that happened very slowly. When you had a score or so of these people – diplomats from various countries residing in Paris, for example, and when governments wanted to rival one another in their display or when very wealthy men were appointed to office, the social life itself became a factor in the case. Ambassadors would gravitate toward one another, exchanging ideas about France, or about their own countries, or about the international situation generally, or about the practice of diplomacy itself. A kind of *esprit de corps* developed among them – they were colleagues in the same profession – in a certain sense came to feel like members of the same club. Not only could every country conduct its foreign affairs with better information – with some knowledge of the very springs of foreign policy in other countries – but the personal relations between various diplomats oiled the wheels of the diplomatic system itself – indeed ambassadors might still be friends though their governments had quarrelled, and the effect of this was to smooth the transaction of business. Furthermore, you now found ambassadors falling in love with the country in which they resided – especially once the social life of diplomats became a brilliant affair. Sometimes they were open to reproach because they became too fond of the people they were living with – too ready to see the other parties' point of view. In the time of Napoleon, the Russian government complained that, whenever they sent an ambassador to Berlin or London or Vienna, the man quickly went native in the country in which he was residing. The complaint was a just one – these men, when their turn of office had expired, too often refused to return home. I once read through a great many of the United States' newspapers for the years 1848 to 1849, and these journals made it their constant complaint that an American couldn't go to Europe as an ambassador without succumbing to the fascination of the courts and kings. When I was young, President Woodrow Wilson once made the remark that if America sent an ambassador to London it only took the man six months to become anglicised. Even in the 20th century, the ambassador abroad has sometimes entered into the ideas of the foreigner more than his own country liked; and before 1914, for example, the British Foreign Office thought that its ambassador in Berlin was too ready to see the German point of view; while the Germans felt that their ambassador in London was too ready to see the English point of view. In reality, all this was very useful – if there does exist a quarrel between Germany and England it is better that the matters be threshed out by diplomats who have been personal friends – far better also to have resident ambassadors always on the spot: ready to discuss the trouble before it is too late. International crises sometimes begin in a small way and something can be gained by catching them at an early stage. If you wait until things become serious and then send a special negotiator from home, he does not have time to feel out the ground or to learn how to handle the foreign negotiator. And the quarrel may have gone too far so that it is difficult to persuade anyone to withdraw.

And the result of all this was that (though nobody had intended it in advance) a change occurred in the character of diplomacy itself, even a change in men's feelings about its underlying purpose. There gradually developed a view that the prime object of the diplomat was not to collect information, not even to assist in the settlement of crises and quarrels – his real job was to keep perpetual contact with a foreign government, to maintain the reputation of his country abroad and above all, to generate confidence: to establish confidential relations. And this last came to be the aspect of the matter on which the diplomatic profession itself laid the greatest stress. Sometimes I even think that in our time they have protested too much about this – as though the real object was to build up a nice feeling between governments or peoples, when also they do of course have to look after the interest of their own nation and to promote the ideals of their own nation. There have been times when there were diplomats who worked hard to build up confidence, but did this only to enable themselves to pull off a confidence trick at the finish. Cardinal Richelieu helped very much to produce a new attitude in all these matters – he insisted that the nations should be perpetually negotiating with one another, ambassadors constantly talking to the foreign secretaries in the countries where they were residing: they should negotiate even when there was no quarrel to be settled, no serious problems to be resolved, no tangible profit to be gained from the discussions. He thought that in this way you would get to know what is happening in the world: but also he had it in mind that you should discuss things for the purpose of getting to know one another better, just for the purpose of building up confidence – rather like a member of parliament in England who goes to nurse his constituency. This man, Richelieu, carried his advice to a point to which the 20th century refused to follow it. He maintained that the government should never stop negotiating even in time of war – somebody once accused him of starting to negotiate for the end of the war before hostilities had actually begun. Apart from all this there might be a case for saying that the economic developments in the world, and particularly the rise of big business affected not only the techniques of government but also the conduct of diplomacy. The habits of transaction, adjustment and bargaining, but also the notions of credit and confidence that are so important in the realm of finance, had a particular relevance in diplomacy.

Of course, once it was realised that a great part of an ambassador's purpose was to build up confidence in his country, there was a special point in choosing respectable and reliable men for this post. It turns out that lying isn't of as much use in diplomacy as one might imagine: if it affects significant matters of business, it is almost bound in the nature of things to be found out very soon and the man himself is unmasked. The untruthfulness tends to appear perhaps in recent centuries in the reckless gossip and small talk, but that is the case also with dons at high table; and typical also of dons is the little habit which one ambassador pointed out: when reporting a conversation to their home government they described themselves as making clever retorts and repartees, which they never made at all, but in retrospective reflection wish they had made. I have been interested to see how in the 19th century, if a diplomat was a liar or did a bit of cheating in a matter of real business, he quickly became a marked man. Hitler and his ambassadors certainly gained some real advantages by barefaced lying, and they were believed because nobody imagined that anybody could be so foolish as to lie in that kind of business. The exposure was bound to come quickly, and you couldn't quite play that trick again because nobody would believe you next time. It becomes enormously difficult to arrange a piece of diplomatic business with a man who can't be trusted, and sometimes any government is liable to find that it has an urgent need to be believed – an urgent need that somebody should credit its promises or assertions.

Then, as quite a separate matter – but while some of these things were already beginning to happen – there occurred a new development altogether, and it produced a more tangible change in the situation. Scholars and writers began to meditate about diplomacy, which now acquired a certain extension and depth through the intrusion of what I might call 'the thinker' – the intrusion of perhaps a little theory, perhaps a little science. It is a curious thing that writers of political treatises, manuals for princes, had devoted a great deal of their space to the art of war but said nothing about diplomacy and foreign policy. A famous Moslem author of this kind, Al-Mulk in 1086[4] had written a work in which he pretended to talk about what we should call ambassadors, but he treats them really as collectors of information and the only technical tip he has to offer is that if an ambassador has something unpleasant to communicate he had better tell the chief minister and let *him* pass it on to the monarch – not confront the sultan with it directly. Machiavelli, in his advice to princes and governments, speaks endlessly about the art of war and stands as a founder of modern military science; but his few maxims of foreign policy are weak and he seems not have realised that diplomacy is a special art, a special field that you could mediate upon and analyze. Towards the end of the 16th century there appeared a particularly fashionable treatise called *Reason of State* (which was defined as meaning the art of government).[5] This not only gave a great deal of its space to military matters – it actually discussed the foundation, the expansion and the downfall of states – but, though it had a page or two about alliances, it dealt only with the ceremonies they involved; it never came near to discussing a problem of diplomacy. From the days of ancient Greece, political theory had concentrated itself on the idea of the state and had treated this as though one state really existed in the world, so that the state was the final thing in human organisation. One idea did emerge in the Renaissance but it was a very elementary one. It was based on the thesis that if a small power is threatened by a greater one, it must set out to collect allies to make up for its deficiency. It is what we call the simple idea of the balance of power – based on the analogy of a pair of scales – a case of piling up a whole lot of little weights on one side until you have counteracted the big weight on the other side.

It would appear to have been towards the end of the 16th century when people seriously began to tumble to the fact that although Machiavelli's political recipes might be sinister, there was still a point to his idea that you cannot escape power politics, since force possesses a sort of unanswerability. It was realised further that Machiavelli had made a genuine contribution when he taught that by quasi-scientific methods you could discover maxims of policy and rules for political actions – you could get some help from history for example – a thing which produced interesting results. In the Netherlands, in the closing decades of the 16th century, students and writers now began to apply this in the realm of foreign policy. But you needn't go to history – you could apply a more scientific mind to the events that were going around you, rather on the lines of observing how things work in politics, and a man like Francis Bacon produced some interesting aphorisms for practicing politicians. A man like Cardinal Richelieu self consciously adopted the method, though Richelieu didn't believe that it was wise to take too much notice of history. In a sense it became curiously analogous to what was taking place in the natural sciences at the time – especially the science of mechanics – which during the same time was at an interesting stage in its development. You had a rather different eye for the way in which things worked in politics – one example would be a famous thesis of Richelieu's which you might think has a special application at the present day: the thesis that, in respect of foreign policy, small states have more freedom of action than Great Powers, and that in the case of an alliance the bigger partner has greater reason to fear desertion than the smaller one. Furthermore, if you once accepted Machiavelli's view about the basic importance of force – seeing in Europe as a whole a general play of forces (all the states exerting push and pull on each other) – if you see Europe as a map of forces, it becomes rather more easy to look at the whole picture with a scientific eye, especially as force is a thing that can more or less be measured, and the 17th century was familiar with diagrams of forces. By the 18th century men were regarding Europe in its entirety as a system. They became highly conscious of it as a 'states system' which made an interlocking group – one historian after another attempted a history of a 'states system' – and they wondered whether there had ever been any real precedent for such a thing in the past. And this was useful – it enabled you to look on the whole of international politics as a system and you could ask yourself what makes it tick? How does this machine work? Suppose we could look at all the states on the globe in 1970 and see them as an interacting system it would greatly increase our ability to get intellectual command of the whole situation. We would gradually come to envisage the European group of states as something analogous to the solar system of Sir Isaac Newton – the states, like the heavenly bodies, exerting a pull on one another – a pull that varied with their size and distance, but held in beautiful equipoise; the situation being such, however, that if one of them were to increase greatly in mass, the rest would have to rearrange their distances – the states of Europe reshuffling their alliances – if the equilibrium were to be maintained.

All this was to have quite an effect on the world – it enabled men to give more structure to their thinking about diplomacy – and it led to curious consequences, some of which had moral implications. So far as I know, it was a French divine, Fénelon, who, at the beginning of the 18th century, first set out the basis for what became the accepted theory, an attempt to put his finger on the thing that made the whole system work. There can be no doubt that he was greatly affected by meditation on what had been the striking experience of 17th century international politics – everybody condemning the Spanish Habsburgs for being such a menace to Europe and then, perhaps rather late in the day, turning around and discovering that the France of Louis XIV had become the real aggressor.

Fénelon made a point which modern history down to the present day has tended to confirm as Spain, France, Germany and Russia have in turn achieved a European predominance, and the United States now a global predominance. He declared that, if a state comes into a position in which it can do what it likes with impunity, it will in fact become an aggressor even if like France, it has hitherto been earning the gratitude of smaller states for saving them (as well as itself) from a previous aggressor. If you object and say that a country might happen to have a virtuous ruler at the moment when it comes to the top of the world, Fénelon is ready with his answer to this. He insisted that the good conduct of a state will last no longer than a single reign. He clearly holds the view that the new situation will bring the wrong kind of person to the helm, the wrong sort of successor to the throne, or that it will generate pressures which the government will be unable to resist. He calls, therefore, for a balance of power – the making or the shifting of alliances to prevent any country from achieving such a position of supremacy. He points out that if you wait until a power has actually reached that summit, the situation is very likely to prove irreversible. He even seems to feel that the prevention of this ought to be the supreme purpose of diplomacy, whereas I think that in the 19th century they would just have held that it was a point which ought never to be lost sight of. In any case, he argued that the object is of such moment that it ought to override the internal laws of any state – override for example a law of succession which might result in a monarch's acquiring a further inheritance and so accumulating too big a combination of territories. Fénelon's view is particularly significant in that it bridges the unbridgeable gap, it relates morality and international politics to the question of the distribution of power. At a later date one sees an extension of this view: the notion that if power were fairly and evenly distributed in Europe – or, if indeed a group of distinctly uneven states were kept in a stable balance – governments would tend to have only limited ambitions. The map would not offer such obvious openings for aggression, such special temptations to a political adventurer, and the general stability would produce a habit of moderation. Each would be too well aware that if it behaved unreasonably, a powerful combination would be formed against it.

By a further course of quasi scientific reflection on the whole 'balance of power system' additional maxims and precepts developed and when you put them together, as they appear in diplomatic documents, parliamentary speeches and literary works they formed a self-consistent code, involving a general attitude to the problem of foreign policy and war. In a sense they were scientific corollaries of the Fénelon system, but also they were put forward as moral requirements; they lead to moralising and the breach of them was morally condemned. They were principles necessary for the preservation of the system, the preservation of international order and it was claimed that there should be loyalty to this international order – in fact some writers almost ran to poetry when they wrote their hymns of praise on behalf of the system as a whole. But the maxims and precepts claimed to be at the same time only a definition of the ultimate good of any state – for every state had an interest in the maintenance of the whole order – a point which even Frederick the Great came to insist upon. So that the whole set of teaching amounted to a demand that the individual states should never lose sight of their long term interests.

These people didn't try to get round the fact that states play for their own advantage, and they didn't pretend that warfare could be eliminated altogether. They demanded that governments should put a curb on their ambitions and that warfare itself should be limited – warfare should be conducted only for limited purposes. It was still true, as it had been previously, that a state which went to war to defend itself or to vindicate its legitimate rights or to defend the weak, would be conducting what was called a 'Just War', while an aggressor would be guilty of an 'Unjust War'. But even a 'Just War' could be carried too far, and one's own legitimate rights could easily be pressed unduly; in any case one must never make it an object to destroy the wicked, one must never turn such a struggle into a 'war for righteousness'. Governments were told that they must state the concrete things that they were fighting for – the particular piece of territory that they meant to acquire for example – for then you knew where you were and if necessary you could make a compromise – a thing which becomes impossible if you take the line that you are out to rid the world of evil. Indeed they argued if a war went on too long, the rights and wrongs of it might change – because you can't go on measuring the justice of it just by thinking of the way it began. Some different power may be responsible for refusing to allow it to end. Repeatedly one is told, therefore, that it is always wrong to fight for the destruction of an enemy – for somebody will be waiting to fill the vacuum that this creates – the way will have been opened for the next aggressor. And as Fénelon said, you yourself, demanding more and more guarantees against the power that is the present menace, may precisely by that method, emerge as the next danger to the continent before anybody quite realises that this is happening. In any case, you must always remember that the enemy of today may be needed as the ally of tomorrow. And as a consequence of all this, peace treaties should always be moderate – always taking into account the need to preserve the stability or the balance of the entire system. One of the excepted principles of this body of teaching is the rule of noninterference in the internal affairs of other countries, and England was particularly attached to this rule, for having had her 17th century revolutions she didn't relish the idea of being faced by a majority of states that were under an absolutist kind of monarchy. But the system went further than this, emphasizing the drawbacks of what we should call 'ideological' warfare, because this was like a reversion to wars of religion – wars so fanatical that they admitted of no compromise. Those who supported Catholicism in the 16th century had ignored the balance of power, desiring total victory for the religion even though it meant carrying Philip II of Spain to complete supremacy. When William Pitt declared war on revolutionary France in 1793, he did it because of her offences against international rights. Charles James Fox, basing himself on the theory I am describing, resisted the idea of turning the affair into a war against the Revolution as such.

These ideas, and the kind of thinking connected with them entered diplomacy and gave it a certain intellectual structure. They helped to shape the mentality of the diplomat – to provide firm ribs for his general outlook – in a sense they created what in 1919 men began to call 'the old diplomacy'. They received their ripest statement among the publicists who, at the beginning of the 19th century, presented their case against Napoleon – the case against having a Europe that was ruled by a single man – something like a repetition of the Roman Empire. These people claimed that it was better to have Europe as a patchwork of smaller states, all based ultimately on a single culture but enriching it by their varied local manifestation of it. Their most important thesis was that the balance of power is the only thing which enables small states to continue in existence in a world so much at the mercy of force – the only thing which enables small states to have an independent foreign policy, instead of turning into mere satellites – the balance of power in fact gives them a role in international affairs – they are the people who may shift the balance. There are occasions between 1700 and 1814 when the precepts of this system were carried out in a signal manner; but there are occasions, too, when the contravention of the precepts was condemned in a signal manner, and these latter are perhaps the more significant. In the 19th century men forgot a lot of the theory and the literary tradition that I've been discussing came to an end, but the maxims and precepts must have entered some subconscious region of the mind – if anything they were better kept in the 19th century than they had been in the previous one. The signal offence against them was by Bismarck, who generally followed and understood them better than anybody else; but, by taking Alsace-Lorraine from France in 1871, he provided the exception that proves the rule – the mistake was fatal to Germany herself. However, it was from the beginning of the war of 1914 that all the powers broke all the rules – and, as had always been predicted, this meant the destruction of the 'states-systems' itself. The maxims and rules for the preservation of an international order that once existed were henceforward at an end.

If you follow Fénelon and say that the supreme object of diplomacy is to prevent the emergence of a predominant power, one unfortunate thing results. Your diplomacy may become too exclusively concerned with the question of the distribution of power. I think it was this that made 18th century diplomacy desiccated and arid sometimes – courts and cabinets weaving their spider webs – almost making it a case of art for art's sake. In the 19th century they were wiser and they seem to have secured what was tantamount to a rectification of the formula. They tended to hold that, whatever your diplomatic object, you couldn't ever afford to lose sight of the question of the distribution of power. In fact whether your object in diplomacy is to promote your countries' interests or your countries' ideals, the rules tell you how far you can go without endangering the whole system. The 19th century saw a further change. You ceased talking about the 'states-system' and you envisaged rather a handful of Great Powers – five or six of them. When you talked about the balance of power you had in mind a balance between just these leading states. A most unexpected thing resulted – The Great Powers in spite of their signal sins, tended to become respectable, their positions giving them not only more heavy commitments, but also a greater sense of responsibility. Their wider territories connected them with a larger range of issues and this in turn induced them to envisage the interests of the whole 'states-system', the whole of Europe. Their leadership provided some resentment but even at the Congress of Vienna in 1814–15, they were showing this wider view of their role – addressing themselves to questions of common interest – the rules governing the diplomatic profession, the Barbary Pirates, the internationalizing of rivers, the slave trade, and the need for a European concert. Sometimes they were anxious to prevent the recklessness of the smaller states – the Balkan states for example – from leading to a war in which they themselves might have to become involved. And even admission to the group of Great Powers ceased to be a question of mere size, mere possession of brute force. It became a sort of status – it was a question of receiving recognition from other members of the club – a question of your being invited to take part in the discussion of general issues. If you were strong and played your cards properly, the other Great Powers would be made to realise that they couldn't afford to leave you out of the main discussions.

In 1919 men had no feeling that an international order had been destroyed through a war that had broken all rules for the maintenance of such a system. They felt on the contrary that, in a 'war for righteousness,' the last seat of evil had been eliminated and now, for the first time in history, an international order had been installed. The cry went up against what was now called 'the old diplomacy', and it was the diplomats who most definitely came into disgrace. The theory was held that they were the creators and begetters of those wars which in reality they had rather failed to prevent. Diplomats don't like wars – first of all because they don't want to fail – secondly because it puts them out of a job (you can't continue as ambassadors in Berlin after your country has gone to war with Germany). Thirdly, war means that the military men came to the front – the diplomats have to take second place. Above all, in 1919, there was a violent opposition to the principle of the balance of power. It was regarded as having been too complacent in its recognition of the factor of power in politics. There was also a feeling that it turned international business into an affair of mean calculations, instead of ranging all problems into a straight issue of right against wrong. The view after 1919 was that there was no need for any balance of power – for if a state took to war or aggression, all the rest of the countries of the world would simply rise in anger against it. I wonder whether the period 1919 to 1939 doesn't represent the low-water mark in the overall of history of European diplomacy.

After 1945 the decisive factor in the case was a point which I am interested to see is being forgotten – for a new generation it is impossible to recapture the experience. It was the way Russia established her predominance and imposed her regime on one country after another in central and Eastern Europe and the Balkans – all the regions that had been emancipated in 1919. For those who watched these movements it was a vastly greater shock, a vastly greater horror, than the Russian invasion of Czechoslovakia in 1968, because in 1968 the enslavement had already continued for twenty years, during which time nobody had believed that there was a possibility of ending it. Nor will the present generation ever understand how we held our breath in the years after 1945, wondering whether the United States would continue her isolation – whether anything could make her realise not only the danger to Europe but also the ulterior danger to herself. In that situation nobody in the period after 1945 could hoax himself with the view that force was no longer a factor in politics or that it could be eliminated from one's thinking merely because one didn't want it to be there – merely because it ought not to be there. In fact the balance of power soon became a tremendous factor behind the calculations of diplomacy again, and the term continually recurs in the discussion of these matters, though it could be held that today there is no international system and that for other reasons, the concept is more difficult to apply. I notice that some of the abstruse theorists of the United States who are trying to make a new attempt at a science of international politics, are attempting to build it on the principle of the balance of power, and they have even reproduced some of the old laws or maxims of the 18th century, though – not realising that they had any predecessors – not realising that there was anything to be learned from history – they have sometimes stated them less adequately and their results have been surprisingly thin.

I personally think that some of the diplomacy of recent decades has been better than at any other period during my lifetime. I remember how the pundits once declared that we should never get through the 1950's without a general war, and I think the world ought to be more grateful than it is for what was achieved during that decade when problems were solved which were far harder nuts for diplomacy to crack than anything that confronted us in July 1914. Another curious fact is that: supposing you were to take the case of the older states – the states of Western Europe – their diplomacy nowadays is much more reasonable than at any earlier period. That is partly because it is a case of diplomacy between what are virtual allies, and though the quarrels between allies can be very grim, the resulting negotiations are always carried on in a better spirit. Also, the western European states don't really have the option of going to war with one another – there are not many weapons that they can use except sweet reasonableness. Above all, they are not faced with the really heartbreaking problems of diplomacy. They don't have to feel that if they play a card wrong, if they lose a trick or sell a pass, the whole game will go to the potential enemy. Here, I imagine, is the agony of being a world power nowadays. In all this I have had in mind the diplomacy that foreign offices conduct with one another. I always imagine that, behind the scenes at UNO – when states are bargaining for the support of one another's proposals – diplomacy may be a sinister as ever it was. But diplomacy is a creative thing and it has had its creative artists – Richelieu, Kaunitz, Vergennes, Metternich and Bismarck for example. Some people have been able by diplomacy to minimise the effects of a great defeat in war. Louis XIV was able by diplomacy to achieve his conquests with a great reduction of actual fighting, a minimum of war; Metternich was the leader and master of Europe for a longer period than anybody else in modern history – and was more effective in that role – at a time when Austria's military position was not at all commensurate with this – he did it by persuasive methods, diplomacy and charm. There are occasions when I think it is true to say that a state has even been able to turn its own weakness into a diplomatic asset – Italy in the late 19th century would be an example.

Notes

1. Written in 1969, the original of this manuscript is in: BC File 137 (eds).
2. Cf O. R. Gurney, *The Hittites* (London 1964), 75–9 (eds).
3. K. W. Schweizer and M. Keens-Soper (eds), *The Art of Diplomacy* (New York, 1983) Appendix I (eds).
4. Nizàm-al-Mulk, *The Book of Government or Rules for Kings*, trans. H. Darke (London, 1960), chs XX, XXI (eds).
5. G. Botero, *Della ragion di stato* (Venice, 1589); English trans. by P. T. and D. P. Waley, *Reason of State* (London, 1956) (eds).

20
The New Diplomacy and Historical Diplomacy[1]

In one or two of the papers which have been presented to this Committee, it has been suggested that the rules of policy which were evolved (and which came to receive recognition) in the days of 'cabinet diplomacy' are not valid in the world of modern democracies.[2]

After the First World War it was easy to imagine that the diplomats were responsible for the catastrophe – responsible for the wars they had been unable to prevent. There occurred, therefore, a strong reaction against the 'old diplomacy', and particularly against what was called 'secret diplomacy'. In any case, it was asserted (sometimes even in the highest quarters) that, in the future, policy would have to be tailored to the needs of modern democracy – conducted under 'simpler' rules than in the 18th and 19th centuries. In 1919 there existed among the Western powers a utopianism which had no parallel after the end of the Second World War.

On the other hand, there were people in the 19th century who, as they contemplated the advent of democracy, already feared that the coming age would be prone to wilfulness in foreign policy. They thought that the masses would give way to passions, moral indignations and short-cut forms of reasoning, lacking the patience for the understanding of 'the other party' (the foreigner), lacking the foresight for the pursuit of long-term objects, and failing to realise the things that might be achieved by diplomatic methods – failing even to see that diplomacy, instead of being a mere game of intrigue and subtlety, might actually be a creative art.

In view of this we may wonder whether the proclamation of a 'new diplomacy' and 'simpler' types of policy in 1919 was not itself an example of the danger which some of our predecessors were dreading – in fact, a facile attempt to pander to the self-esteem of the masses. The call for a 'simpler' diplomacy envisaged the policy of the 'good' clearly directed against the 'bad'; and it involved just the inflexible kind of self-righteousness, and the unhistorical attitude towards the past, which might be expected to characterise an age of young democracies. It might have been better, after 1919, therefore, not to humour the inexperienced democracies so much, but to press upon them the urgent need for education in the whole problem of international affairs, now that they were taking upon themselves such great responsibilities in these matters. Parliaments themselves (at least in their better days) had shown a kind of self-discipline in this field, and in many parts of Europe had allowed the royal prerogative a special authority in foreign policy. Perhaps, instead of proclaiming so complete a break in 1919, it would have been better to take the line that here was the moment for asserting (and insisting upon) the continuity of history, or the importance of gaining every possible benefit from man's long-term experience. In any case, it is possible that statesmanship and rules of policy are not amenable to the kind of arbitrary re-definition that was envisaged in the years after 1919 – not capable of being rendered 'simpler' by an act of will, or by the mere fact that democracies needed to have them signified.

In favour of this critical attitude towards the developments which took place after 1919, it might be argued that, if there are rules of diplomacy and laws of foreign policy, these must be valid whether the business is conducted by men or women, whites or blacks, monarchies or democracies, cabinets or parliaments. If it is unwise to exploit a victory over-much, or to forget that the enemy of today may be needed as an ally tomorrow – if it is wrong, through reliance upon the virtue of a certain power, to allow that power to get into a position in which it can misbehave with impunity – these things do not become more admissible when practiced by democracies rather than by monarchies. It may be true that some of the routines of diplomacy – some of the techniques and detailed practices – may depend on conditions (on the state of communications, for example, or the character of the *régimes* involved); but this can hardly be the case with rules of policy and the way in which consequences proceed out of causes in international relations. Out of the experience of centuries, there ought to have arisen, if not something like a science of diplomacy, at least a ripe kind of wisdom in regard to the conduct of foreign policy – rules or maxims possessing a permanent validity, at any rate so long as policy is being operated within a system of nation-states. After 1919 the principle of 'the balance of power' was treated as a kind of prescription – a policy which it was open to one to adopt or to reject. In fact it is rather a formulation of the way in which national interests do actually operate, and already operated before the principle came to its conscious formulation. If the principle of the balance of power was useful or valid in the 18th century, it was likely to be useful or valid in the 20th century, for its *rationale* never had anything to do with the character of particular *régimes*. In reality the principle first came to consciousness not even in the world of nation-states at all, but among the city-states of the Renaissance.

We may ask therefore: Is there not something which can be condensed from the experience of many centuries, and is there not a point in extracting or collecting this, and turning it into a teachable form? Or are we to say that a new kind of *régime* must face the additional difficulty of inventing a new type of diplomacy? The Napoleonic Empire represented a new type of regime, but it is remarkable to see how much Napoleon learned of the fundamental rules of policy from a conscious use of history – a use only feasible perhaps for a man of peculiar insight. Also, the lines of his foreign policy projected themselves in quite an astonishing way from the grooves that had been formed by traditional French diplomacy. Something of the same would appear to have been true in the development of the policies of Soviet Russia. In a world where so many powers, great and small, are continually pulling and thrusting, are there no rules about the play of forces – about the way in which consequences proceed out of causes – that remain fundamentally the same, whatever further complications may have been superimposed by the advent of a new regime? The question is relevant to our Committee which is interested in the problem of the possibility of a 'theory of diplomacy'. It prompts the further query: Whether the 20th century has done justice to the permanent and fundamental elements in the structure of what we call a foreign policy – whether it would not have developed faster and further if, instead of emphasising the break with the past in 1919, it had set out to retain the accumulated wisdom and experience, and build its superstructure on the top of that. The new democracies have had to learn at great cost the consequences of the breach of certain maxims the importance of which our predecessors had already deduced from bitter experiences. The democracies even failed to acquire the necessary knowledge of the nature of diplomacy itself, remaining unaware, for example, of the need for armaments under the conditions of the inter-war period – the need for armaments not for the purpose of war-making but for the purpose of e.g. securing the necessary 'pull' in negotiations.

In this connection, it may not be entirely irrelevant (and at least it is not irrelevant to the purposes of our Committee) to take note of the very considerable decline in the study of diplomatic history which has been taking place in recent decades in this country. It is partly an incident in the general decline of what we customarily call political history.[3] It is partly due to the fact that our 'general history' (i.e. our European history), instead of being a dry account of the external relations of states has been turning into a survey of the development of our whole civilization. The decline is most remarkable of all among research students, however, and here it is particularly regrettable; for diplomatic history, owing to the nature of its documentation and to the kind of precision-work it requires, happens to be an excellent initial training for an historian. A conscious reaction against diplomatic history in general began to occur in the latter half of the 1930s, and this was partly due to the influence of Marxist history and the tendency to reduce policy to certain economic determinants. A particular campaign was waged against the kind of history that was written out of diplomatic documents, on the ground that the actions and policies of states were based on considerations not recorded in such documents. Diplomatic documents had been the first to occupy the historian after the rise of a more scientific history and the gradual opening of the archives in the 19th century. They acquired such status that, pending the release of further types of documentation; diplomatic reports were used as the authorities for the *internal* histories of countries in a way that would not be admissible at the present time. Diplomatic history was an old subject, therefore, and by the middle of the 1930s research students were pressing upon the date-limits of our archives. Whereas in 1919 these students were grouped in the period around the Congress of Vienna, by 1934 there were congested around the period 1875 – 77, and many of them were finding it difficult to discover untrodden ground. Economic history, on the other hand, was a comparatively new subject, and Marxism had important effects even on non-Marxist historians in the latter half of the 1930s. A further reason for the decline of diplomatic history was the fact that its writers were sometimes too dry and austere, producing a history that was too much a mere digest of the documents, without sufficient reflection on policy or its attendant circumstances – without sufficient thought about the nature of diplomacy as such. In my young days, students of European (as distinguished from British) history often received their introduction to the study of a foreign country through an initial piece of research in diplomatic history. They could find at least one important section of the necessary documents in our own Foreign Office archives, and they could gradually acquire experience of a foreign country, whereas the internal history of that country would have been more appropriate to its own native students. At the present day there is a serious decline of European History in our universities, and it has become difficult for us to fill chairs (either here or in Commonwealth countries that have approached us) where a 'European' historian is required. (If a chair in European history is now vacant the four most obvious candidates for it in this country would be refugees who have settled here).

Nothing contributes so much to rescuing a University History School from insularity as a strong department or a strong representative of European history. Certain disadvantages seem to spring from the fact that military history and naval history are almost entirely neglected in practically all our universities. Hardly more than one or two universities in Great Britain have a department of International Relations. There is a danger that the country at large may lack the nucleus of people which it certainly had until 1914 – people, outside the circle of government, who are equipped to understand something of the real nature of a foreign policy. For this reason the present question brings to my mind the further problem: Whether the character of 20th-century diplomacy has really rendered the historical study of the subject less relevant than it was forty years ago?

II

In one sense the reaction which took place in 1919 against the 'old diplomacy' may have had a certain amount of justification. There was considerable subtlety in much of the diplomacy of the eighteenth century, as though cabinets could weave their spider webs, almost pursuing their policies as one pursues 'art for art's sake'. In reality, however, it is probably true that the arguments which have usually been made from this fact require to be completely reversed. We ought not to infer that the 20th century presents us with a special case, to which the long-term experience of the past cannot apply. It is the 18th century which is the special case, being handicapped for the conduct of diplomacy in a way that neither the Renaissance city-states nor the 20th-century nation-states were handicapped.

In the monarchies of the 18th century the modern idea of the state was as yet unachieved. The interest of the king, as in a certain sense the owner of the property, was sometimes to be distinguished from the interests of the country itself, so that 'dynastic' policy sometimes run counter to 'public' policy. In the first half of the century in particular, two systems of policy might be open to a given government, and on many occasions these two systems were simultaneously pursued. In the case of the Hanoverian Kings of England, in the Spain of Elizabeth Farnese, in the France of Louis XV and, later, even in Russia, the monarch could actually be at feud with his own foreign minister, or could conduct a second diplomatic system, almost as a kind of palace-intrigue. The atmosphere of melodrama and the flavor of conspiracy which attend the conclusion of the famous Anglo-French alliance of 1717 (and surround the Diplomatic Revolution of 1756) are the result of an anomalous system in which secret agents, political adventurers and adroit manipulators flourished. Here is the primary reason for the excessive subtlety of much of the diplomatic manoeuvring of the 18th century. Many of the features which diplomacy came to acquire as a result of this situation must be regarded as incidental to the situation itself, and as unconnected with that inheritance of long-term experience with which the present paper has so far been concerned. But in view of this fact, it is remarkable that the 18th century did at the same time appropriate so much of the world's past experience, and consciously reflected upon this experience, producing comments and inferences that possibly have permanent value.

On the other hand, it is not clear that even the anomalies of the 18th-century situation do not have their subtle parallels in our own day. To an outsider, at least, it seems possible that such changes as have taken place in the conduct of foreign affairs in our time have not been in the direction that the men of 1919 so blithely expected. In so far as the coming of democracy has added a new factor that has to be taken into account – an extra ball to keep in the air – it may have rendered the task of a diplomat more complicated than before. A Foreign Office that has to steer amid the vagaries of public opinion may have analogies with the public minister of the 18th century who had to make himself palatable to a court. In so far as 'open diplomacy' came to mean 'diplomacy by conference' it must have entailed certain changes in technique and procedure – perhaps the introduction of new practical devices. But certainly it must have had disadvantages as well as advantages, and one wonders whether the success of conferences must not ultimately depend on new forms of what used to be called 'secret diplomacy'. The development of communications must have influenced the conduct of foreign policy almost as radically as the coming of democracy. In some respects the machinery must have worked very differently, (e.g. in the Napoleonic age) when a diplomat abroad sometimes had to wait for months before receiving an answer to a request for instructions. Perhaps the speeding-up of communications and the advent of democracy have combined, therefore, to make the ambassador less significant in both the manufacture and the execution of policy. Together, they have both produced the need and provided the facilities for frequent direct meetings between Foreign Ministers or Heads of States. On the other hand, the effect of ambassadors in the days when they had an important independent influence was a qualifying or complicating one, tending to cushion or mediate or mitigate the consequences of a policy or the conflict between national interests. Without this complicating factor, the play of forces is more direct – the operation of national interests becomes all the more clear and clean. Policy is not more free of whatever rules or laws may permanently govern its: it is subject to them in a more unqualified manner than before.

Since 1919 it is possible that there has been a further change or development in technique. A certain move may be made in foreign policy not so much for the sake of its immediate diplomatic effect, but rather with an eye to the impression that it will make on public opinion either at home or abroad. In a converse manner, one suspects that certain lines of propaganda, adopted at home, really envisage some negotiation with a foreign power – that is to say, even an article in a newspaper or a question in the House of Commons may be a move in the diplomatic game. One can hardly believe that the effect of this is to make the work of the diplomat less complicated than it was in the days of our fathers. I am not clear that it is more easy for the members of a democracy to understand movements in diplomacy today than it would have been (if they had had the information) under the conditions of the year 1913. Indeed, fundamentally, modern democracies may be more at the mercy of their own governments or foreign offices (where these latter are strong-minded enough to desire and to take the lead) than the Liberals in the time of Sir Edward Grey. One of the serious effects of the coming of democracy may be the way in which foreign policy may become a party matter; and this, whether right or wrong, must sometimes weaken very seriously the diplomacy of the country concerned.

But neither this nor the other well-known effects of democracy on foreign policy can be regarded as really new. The changes that have taken place are only a prolongation of tendencies that were visible long before the outbreak of the War of 1914. The attachment to the 'conference' method (and to direct meetings between Foreign Ministers) is to be seen already in the years after 1815. Movements in public opinion were factors of which statesmen in the 18th as well as the 19th century were often compelled to take account. The public opinion in question would correspond with what was the politically effective part of the nation; but Walpole was overborne by e.g. the commercial interests in England in 1739, and was forced to enter upon a war that he wished to avoid. Even Cardinal Fleury was similarly overborne (and forced to go to war with Austria) in 1740 by a party of younger people at the court of Versailles.

It has been argued that the principle of the balance of power is difficult to reconcile with the facts of the twentieth century, when the 'rights' of democratic peoples have come to receive such thorough recognition. In the 18th century, however, a parallel system of 'rights' did exist and received similar recognition; and the operation of this was bound to be very much the same, even though the nature of the 'right' was different (i.e. was dynastic in character). Diplomacy had to be conducted within the limiting terms that were set for it by this system of recognized 'rights'; though if there was any juggling it was with genealogical tables rather than with historical or ethnographical claims, such as have been so familiar in more recent times. In both cases the balance of power, for example, had to be engineered with due regard to the accepted theory of rights. Where these were contravened, it was held, in the 18th century as well as in the 20th (e.g. in the case of a Partition of Poland in both centuries) that a great crime had been committed against the whole order.

One radical suggestion has been made about the present state of Western diplomacy and it is perhaps interesting because its very radicalism springs from the use which it makes of an historical analogy. In the centuries before 1914, views of European diplomacy were greatly affected by the fact that international relations were supposed to be conducted within a 'states-system', and, though the system was virtually a European one, Heeren, 150 years ago, was contemplating its possible extension to all the countries of the world. The states within the system (however bitter their rivalries) belonged to a recognisable cultural group, and were like members of the same 'club' – ready to compete for position within the 'club', but anxious not to destroy the 'club' itself. On the other hand, there had always been a realisation of the existence of countries and powers that lay entirely outside the recognised states-system. The Greek city-states had treated the inhabitants of such regions as mere barbarians, while the Christians of the Middle Ages had also been compelled to fight the total outsider – the power of Islam in particular. It is sometimes held that the conflict between the democratic West and the communist East creates the necessity for this more drastic kind of diplomacy – that here is a kind of super-rivalry or super-warfare, comparable to the struggle of Civilisation against Barbarism, Christianity against Paganism. A similar situation may have existed during the conflict with Hitler, who may have won some of his initial vicotires because the other powers were expecting that he would play the game according to the rules of the 'club'.

It is doubtful whether this whole attitude implies anything more than a counsel of desperation – an insistence, for example, that communism shall not be included in the international order to which we belong. It is always a temptation to believe that the enemy of the moment ought to be cast out of the whole system, and during the First World War we became accustomed to regarding Germany as the single Great Power unworthy to be accepted as part of the civilised order. In reality the analogies in history are liable to be overstressed for the purpose of making desperation-policies more plausible. In point of fact, the Christians did extend their conception of law to embrace the case of warfare against infidels and barbarians. The Turks may have stood outside the order of Christendom, and till nearly the close of the 17th century it was possible to form Holy Leagues against them. From the time of the Renaissance, however, they formed part of the regular system of European alliances, though their handling required certain peculiar techniques. In the 18th and 19th centuries their preservation came to be regarded as essential for the European balance of power. In the 16th and 17th centuries the Protestants and Catholics would have liked to annihilate one another; but diplomacy naturally operated in favour of a new and creative achievement – an international order that comprehended both. Russia today is a member of UNO; that is to say, she too is within the system even while she is working in favour of a counter-system. So long as actual war is avoided, the diplomacy of the 20th century can hardly escape its traditional role – the preservation of an international order, or the further development of such form of order as already exists. If in our diplomacy we are still working for peace (and not merely strengthening our own position of what is regarded as the inevitable eventuality of war), we are *ipso facto* working for the development of an overall 'international order' and for the achievement of a kind of stability. In this sense, 20th-century diplomacy can hardly help having continuity with the past.

The most remarkable feature of the kind of diplomacy that has developed since the end of the Second Word War is one which represents the converse of the suggestion that the issue lies between the representatives of Civilisation and the representatives of Barbarism. Recent years have seen an important development of the role of what might be called the 'moral factor' in international relations; and it is strange that in this period, when armaments have become formidable beyond all precedent, actual weapons have lost some of their relevance and power, and an imponderable factor has acquired unusual importance. It is this which is altering (or which ought to alter) the character of 20th-century diplomacy more than anything else.

The dread of actual war has not put an end to the desire for changes in the *status quo* but has meant that governments feel compelled to seek victories other than the victories of brute force. The existence of so many undeveloped and uncommitted countries (or of regions under various forms of imperialistic control) now means that a tremendous conflict for 'influence' is taking place in time of peace. The Marxists (developing the methods of the French Revolution) had long ago brought propaganda and diplomacy to a closer fusion (and at a higher degree of temperature) than ever before. That fusion has now become a momentous matter for both East and West, and has enhanced the importance of the moral factor in international relations everywhere.

The change would not be so remarkable if it took place at UNO, where one might have expected it. But in reality, the remarkable thing about UNO may be the way in which so much of what might be called the 'old diplomacy' underlies its proceedings or decides the grouping of votes. It is rather through the traditional diplomatic channels and in the ordinary relations between governments that the introduction of a new factor seems to be producing something like a qualitative change. In this new kind of world the Egyptians can secure a victory that has no correlation with the power that they actually possess, and they can proceed to make changes in the *status quo* which the West may fail to check, or can only check if they enlist the aid of the moral factor themselves. It has been apparent for some time that Khrushchev is determined to secure victories of this moral (or quasi-moral) kind; and, just as Acton seemed to do, he realises that the moral factor tends in reality to make for changes in the *status quo*. [In fact it has become more powerful than 'legalism' – than those rights (or that *status quo*) which could claim to be established in international law]. But Khrushchev seems to see also that, by means of this moral factor, he has a possibility of producing important changes in the distribution of power in the world. The episode of the spy-plane makes one wonder whether the dramatic use of this kind of sensation may not be capable of having effects comparable to those of a victory in battle. In the case of the Suez episode the moral factor seemed stronger than the claims of friendship, alliance and even immediate self-interest, overriding all purposes shorter than the ultimate and long-term purpose, i.e. the aim of acquiring the confidence of the uncommitted areas of the globe. Sheer accumulated power has not lost its relevance, and violent declamation is one of the remarkable features of the game as it is now played. But an imponderable factor has acquired an importance that is possibly unprecedented; and, so long as the colossal giants confront one another with fabulous weapons which, humanly speaking, they cannot use and do not intend to use (and which they will certainly not find it practicable to use for any peripheral or doubtful purpose) this imponderable factor may operate with overriding effect. There could possibly be a great shifting of power, while the great States most adversely affected by this might not be allowed an opportunity for firing a gun. It is precisely because the guns cannot be brought into play that violent anger, thunderous denunciation and apparent willfulness may have a significant place in a calculated strategy. In other words, the situation may entail (or require) a change in the character of diplomacy – though even this change would not constitute an absolute revolution.

This leads me to wonder in what way operations at UNO may have required a new kind of diplomacy, or resulted in the development of new techniques.

Notes

1. Delivered on 24 September 1960. Original in BC File 329 (eds).
2. Butterfield is probably referring to the papers by Michael Howard, 'Scientific Development and International Relations', presented in January 1960, and the one by Desmond Williams, entitled 'The Role of National Prejudice in the Formation of Foreign Policy', given at the same session (cf Vigezzi, op.cit., p. 328) (eds).
3. As elaborated upon in Chapter 1, Part I of this volume (eds).

21
The Changing Moral Framework[1]

It might be useful to enquire first what were the moral aspects of that system, associated with the principle of the balance of power, which went on existing in an attenuated form down to 1914, providing the basis for what came to be called 'the old diplomacy'. This might have the advantage of showing what a 'states-system' can imply, as well as possibly providing a norm by means of which we can measure progress or decline in the subsequent period.

The theorists of this inherited system had grounded their ultimate defence of it upon the thesis that the balance of power provided the only means by which small states could be kept in existence and could carry on an independent foreign policy (even performing a role of their own in foreign affairs) in a world that was so much at the mercy of force. In other words, those who chose to ignore the distribution of forces would in reality be releasing the world to a predominant power – a Louis XIV or a Napoleon; and the result would be the elimination of smaller states, the creation in fact of another Roman Empire, imposing uniformity on everybody, which could not be the ideal for a lover of liberty. It was far better to have a patchwork Europe, a society of states which, though based on a common culture, enriched that culture by the variety of their local manifestations of it.

The roots of the system went back to another thesis which had been stated at the beginning of the 18th century and which carried moral implications. Fénelon had put the case that if any state were allowed to reach such a predominance that it could do what it liked with impunity, it would in fact become an aggressor, however virtuous it might hitherto have been. Since such a predominance was likely to be irreversible once it had really been allowed to take place, the anticipation and prevention of such a thing were very high – if not actually the supreme – objects of foreign policy. In all this the unbridgeable gap seemed to have been bridged and some sort of relationship established between morality and the distribution of power. At a later stage some people were disposed to believe that if all states were fairly equal in strength – or if a society of distinctly unequal states were kept in stable balance – all governments would tend to become more moderate in their ambitions and would even acquire a habit of moderation. There would be fewer openings for an aggressor, fewer invitations to political adventure; and cupidity itself would be curbed by the constant awareness of the hostile combination that its excesses were likely to provoke.

If international politics were based on these principles as they were initially formulated, the result might be to produce foreign policies too exclusively devoted to the question of the distribution of power. And perhaps this explains why 18th-century diplomacy can sometimes seem desiccated and arid, spinning its spider webs and almost functioning as a case of art for art's sake. In the assumptions behind 19th-century international politics there may have been two subtle amendments in the formulation of Fénelon's basic theses. It was felt that if a power – now France perhaps, now Russia, now Germany – were to acquire an undue predominance, it might become a danger to others; and, whether it did so or not – whether its existing rulers were virtuous or not – one must act on the assumption that one could not afford to take the risk. Therefore it must still be a maxim that one should check the march of any country to supremacy. And, where Fénelon may have thought that the prevention of such a thing ought to be the supreme object of diplomacy, the 19th century would possibly have said, rather, that this point – and, therefore, the whole question of the distribution of power – was a thing which at least no state, in the conduct of its foreign policy, could afford entirely to overlook. In the 19th century, therefore, the rules and precepts of the system that we have to consider may have slightly altered their status. They might not prescribe to a government the ordinary objectives of foreign policy, but they established limitations. They told a state that line beyond which it could not safely go, whether in promoting its own interests or in pursuing its ideals.

There had been emerging in the days of Fénelon the notion – and the feeling – that Europe was a society of states, and that, behind all the conflicts, the nations were somewhat like members of the same club. Without this way of looking at things, there could hardly have arisen the more modern, more complicated, idea of the balance of power as something more than a pair of scales – Europe being envisaged on the pattern of the Newtonian solar system, the various countries kept in equipoise by the pulls they exerted on one another, these pulls themselves varying according to the mass of the various units, varying also (inversely) according to the distance between them. Some writers almost broke into poetry as they sang hymns of praise to the European 'states-systems'; and historians set out to write the history of this system, asking sometimes whether there had ever been a precedent for such a thing in the past or whether its existence was one unique only to a special combination of circumstances. In a sense, the rules that we have to consider were rules for the preservation of such a 'states-system', granted that anything of the sort existed already. They raise the question whether, in a world where these rules cannot possibly be obeyed, a genuine 'states-system', or indeed any fertile and operative 'international order' really exists. From this point of view, the theory, its assumptions and its rules or laws might be compared with what Morton A. Kaplan, arguing from first principles and ignoring the history and literature, deduced in 1954 for a balance-of-power system in his *System and Process in International Politics*.

There was no pretence that the potential aggressiveness of states – large or small – could be eliminated or that war could be entirely abolished; and in a sense the idealism in the theory was that of practical men who, when an evil cannot be entirely removed, seek at least to hold it within limits. The fundamental prescription was loyalty to the 'states-system' itself – a 'states-system' which included both monarchies and republics, both Catholic and Protestant units, and which (rather than any religious party) should be for conductors of foreign policy, the single object transcending the interests of their own country. Such loyalty was enjoined as a matter of duty, but it was also envisaged as a point of enlightened self-interest, since all had reason to fear the overthrow of the system itself. Indeed, the idea was to work out in a scientific manner – very much on what might be called Baconian methods – the way in which that particular order of things could be made to endure and to work to the ultimate benefit of all. On this basis the effect of the injunctions was to require states to work not merely for short-term ends but also for remoter, long-term purposes – indeed, to recognise a more enlightened form of self-interest.

The maxims and precepts have a curious coherency as a kind of programme, and they had currency in the practical world, they had more influence on the practical world than the teaching of those who wanted Utopia very quickly. Yet on the whole – and certainly where the distinction is relevant – they were injunctions addressed to victor-powers, the assumption being that it is not the defeated who are likely to have the relevant options open to them. One might wonder whether they might not have special applicability in the case of victor-powers which desired to establish a 'states-system' or an international order in a world where no such thing existed or such a system had broken down.

States were enjoined to remember that, whatever hope they might have of securing the ascendancy themselves, they always had a profounder motive for avoiding conduct that might bring another of their number into that position. They were urged therefore to put a curb on their ambitions and in particular to restrict themselves to limited warfare – warfare at any rate only for limited ends. It was still true, as in earlier times, that a state which went to war either to defend itself or to rectify an infringement of rights or to assist the weak would be fighting a 'Just War', while those that committed aggressions were guilty of 'Unjust Wars'. But even the 'Just War' could be carried too far if one turned it into a cosmic 'War for Righteousness'; and if the struggle went on too long the rights and wrongs of it might change, the justice of the war being measured not merely from the way in which it began but by the kind of world which would issue as a result of it. One might in fact become the real culprit by simply being the state that refused to end a conflict. Governments were told that they should state the concrete thing that they were fighting for – the precise piece of territory that would change hands at the finish, for example – because here was something that might be amenable to compromise, whereas no compromise was possible if one fought to establish righteousness in the world. It was wrong in any case to fight for the destruction of an enemy – somebody would be lying in wait to fill the vacuum and the way would be opened for the emergence of another aggressor. Indeed, as Fénelon said, you yourself, demanding securities against a present aggressor, and then more and more securities, might advance to the predominant position before the world was quite aware of what was happening. In any case, you might remember that the enemy of today may be needed as an ally tomorrow, some friend of today being liable to emerge as the new aggressor. And, similarly, peace-treaties should be moderate in character, so that the stability of the entire system might be preserved, frontiers themselves coming to be accepted as settled and habitual. One of the older accepted principles was the rule that countries should not interfere in one another's internal affairs, and England precisely because she was more advanced in her liberalism, was particularly attached to this. But the theorists went further than this and insisted that anything like an 'ideological' conflict was almost bound to wreck the very basis of the 'states-system' itself. Those who had subordinated foreign policy to Catholicism in the 16th century had scorned the balance of power, desiring victory for their religion though it meant carrying Philip II of Spain to complete supremacy. A 'states-system' had only become possible when it could comprehend both Catholics and Protestants (just as it comprehended monarchies and republics, despotisms and constitutional states) religion being then abstracted from diplomacy and war, and left to the volition of states or to the realm of reason and persuasion.

In the period down to 1815, policy, and the publicity attendant upon it, would often conform to these principles, and proposals of diplomatic action or recommendations concerning war and peace would make reference to them. Perhaps it is a matter of greater significance that when there was a contravention of this general code, there would sometimes appear signal reproof. Possibly between 1815 and 1914 the contraventions were fewer than in the preceding century, though Bismarck, when he took Alsace-Lorraine in 1871, may have provided the exception that proves the rule, the exception that Germany herself was to have reason to regret. Yet in the 19th century it was as though the rules had sunk into some realm of the subconscious, for the body of teaching seems no longer to have been explicitly formulated and the literary tradition in this field may have come to an end. Also, one ceased to hear of the European 'states-system' and the stage was dominated by what seemed on occasion to be more like a club of Great Powers, the balance itself now being regarded as operative just between these, and the plans of a closer 'Concert' having reference really to them.

What can hardly be doubted is the fact that in the First World War all the rules of the European 'states-system' were broken at once; and though there might be a strong case for saying that the belligerent powers had no choice in this matter, we might wonder whether the predicament would have been quite the same if all the powers concerned had properly remembered and assimilated the theory of the 'states-system' and the urgency of its accompanying rules. The intransigence of the First World War and the drastic nature of the subsequent treaty – whether inevitable or not – had the effect that had long been predicted. They finally broke the old 'states-system' and produced for diplomats those far more radical problems that naturally occur when no such supra-national system exists. It would appear that if a 'states-system' is destroyed through the breach of these rules, a world is produced in which one cannot apply the rules however earnestly one may desire to do so – they merely come to appear irrelevant.

Yet the rules only prescribed a kind of statesmanship and insisted on the virtue of moderation, setting out to curb in particular that national self-righteousness which in international affairs is even a more intractable problem than that of national greed. In a world structurally different from ours, the theorists of the system so feared the effects of passion that they preached a doctrine which can only appear strange to us recommending that popular fervour and moral indignation should not be aroused in time of war, lest when statesmen wanted a compromise peace, they should be thwarted by the public opinion of their own country.

II

If we are thinking of moral factors which are not merely changing principles (like the thesis that war is wrong) but, by a certain effectiveness in their operation, tend on the whole to form a framework for the conduct of international politics, the year 1919 produces a clear break in the story, and there are many things that help to explain the break. The conflict which had now ended had been fought as a 'War for Righteousness', almost as a struggle against the last remaining seat of evil in the world. There was a feeling that the right had been victorious; the world was now 'safe for democracy'; and a great new era was thought to be opening. The Russian Revolution had eliminated the danger that John Morley had had in mind when he asked what it was proposed to do about Russia after Germany had been dealt with. A radically new Europe had been brought into existence, established frontiers overthrown, and the three great European empires – all regarded as sinister in a way – first broken and then gravely depleted. The emancipation of a number of peoples hitherto submerged by the Hapsburg dynasty or by Russia – the rise of new national governments in Czechoslovakia and Hungary, in Poland and the Baltic States – had opened what looked like a honeymoon period for small states – those smaller states which, after 1945, were to fall under a despotism more complete, if not possibly more permanent, than ever before. There was hope among the peoples that the League of Nations had now provided the machinery by means of which the evil of war might in future be avoided. It was even felt that a Europe so largely composed of democracies would be rescued from the warfare that the monarchs had always been so ready to renew.

All this was mixed, however, with a great resentment, a special bitterness against that older order of things which had culminated in so terrible a catastrophe. So completely had history been forgotten that it was precisely the system of the whole preceding period which now came to be described as 'international anarchy'. There was a determination, therefore, to have a break with the past – indeed a reaction against it – and the cry went up particularly against what was called the 'old diplomacy'. The diplomats themselves came to be disgraced, for they were identified with the system – regarded sometimes as the creators or begetters of those wars which it had been their function rather to prevent. Above all, there was a violent opposition to the principle of the balance of power; and perhaps the notion of a balance was difficult to apply to the new situation, where, in one sense or another, those that were not victor-states had almost been removed from power-politics. It was now held that the whole theory of the balance of power had been too complacent in its recognition of the factor of power politics. In any case it was felt that the older systems had had the effect of turning foreign policy too much into an affair of mean calculations. Also that system had deprecated too greatly the idea of the straight opposition of right and wrong in international politics. If any government in future sought to advance itself by sheer force, there would be no mere question of tilting the balance against it. The whole rest of the world would rise in anger against such obvious wickedness. And in this, it was held, lay in point of fact a far better guarantee of future peace than had ever existed before.

All this was the prelude to one of the most unfortunate periods in the history of modern diplomacy.

In any country and any period there will be idealists who ignore the limiting conditions and tend to be in almost standing opposition to the practitioners of international politics. They may contribute therefore to the raising of standards in the long run rather than to the solution of immediate problems. Since such men tend to ask or expect the rest of the world to be better than they really are, they ought perhaps to join that excellent and still more important class of people who, as preachers, teachers or prophets, devote themselves to the higher problem of improving the quality of human nature. For they fail as statesmen – they are not the people to save us from a Hitler or a Stalin when such giants stand solidly in power. Because of the claims that he makes for himself, any statesman – any advisor of statesmen – has the responsibility of rescuing his country or saving the world, in spite of the recognised limitations of human nature. He is not entitled to say that he could have succeeded if he had had to face a different problem or if his predecessors had not spoiled the field or if his fellow countrymen had been less egotistical. He is certainly at his greatest and most magnificent when, by inspiring leadership, he can on a special occasion raise his countrymen above themselves. But the idealist is not coping with the problem if he merely exclaims at the end of the argument: 'Well, human nature *must* improve, must rise to *this* level of conduct – or else . . .'.

Because of the contradictions and paradoxes involved, the realm of international relations, more than any other, is calculated to suffer at one and the same time from the cupidities of the wicked, the anxieties of the strong, and the unwisdom of the virtuous. It is a field in which the self-righteousness of nations can be even more deadly than the phenomenon of national greed. And because the problems and paradoxes of human relations show such constancies of fundamental pattern throughout the ages, it is the realm where man ought to learn most (though in reality he most refuses to learn) from the accumulated experience of the human race. It is surprising to note how many of the faults of the inter-war period had been predicted in earlier generations as the likely tendencies of new democracies, or new classes brought to the seat of power.

Nevertheless, there are moral factors – there are concrete policy precepts arising out of ethical considerations – which do have a certain efficacy in the practical world, whether because of their peculiar relevancy or because there lies behind them a body of opinion that the statesman cannot ignore. They are things which powerful states might find it prudent to adopt, in spite of inconveniences, or would use as the basis for reproaches against a rival that contravened them. It would be difficult to say that between 1919 and 1939 there existed a framework of moral ideas within which international politics normally had to be conducted; but for a great part of Europe, the principle of the self-determination of peoples, the notion of a League of Nations acting from a presidential position, and, further, a programme of disarmament, (such as was taken for granted in the treaty of Versailles) helped to give the period its character. At least a certain amount of lip-service to these principles or ideals was generally required.

However, irrespective of the giant power or the truculent rebel that in any age of history may contravene the general rules or even ignore the claims of morality, there were faults in the very governments that had some attachment to these general objectives. Irrespective of the peculiar difficulties which the special problem of general disarmament still presents, there were discrepancies and contradictions within the ideals themselves, or at least in the way they were apprehended. The principle of the self-determination of peoples has often been difficult to implement. There are regions where peoples and nationalities overlap. Compromises have to be made. There then arises the further problem of the treatment of minorities. It is not always easy in practice to know how far to carry the principle. In the British Isles there has been the problem of Ulster. Well-meaning people have had differences of opinion on the subject of Biafra. It would not be easy to formulate a law (one so clear that it might stand as a moral imperative) concerning the regional limits that should be put to a population's right of secession. An area like Palestine can present problems of considerable complexity. There may even be obstructions to the discovery of the will of a people if the presence or the neighborhood of a great power is liable to affect a vote. There were other complications even in 1919 when, for example, De Valera in Chicago vainly called upon Britain and her allies to apply to Ireland the principle on which they claimed to have fought the war and established the peace. In any case the principle was not applied to India and to colonies in Africa etc.; and the whole international world, including the League of Nations, was prohibited from taking up the cause of these peoples by the somewhat legalistic consideration that this would mean interference in the internal affairs of what was still virtually an empire. A world had been created in which it would be illegitimate to challenge the states that had already filled their bag with booty. On the other hand, the British were able to take the line that they were guiding to a regime of liberty a number of peoples who as yet were unprepared for it; and, in spite of the vested interests involved – in spite of the fact that progress ultimately turned out to have been too slow – it must not be imagined that this was insincere or that the servants of empire, however limited in imagination, were entirely without a sense of mission.

The League of Nations was a body to which not only the United States but also, at various times, considerable powers, like Germany, Italy and Russia, had an anomalous relationship. Many of the individuals most ardent in its support tended to value it as the instrument by means of which the extremity of war might be easily and cheaply avoided. When the League failed to cope with these ultimate problems, it was partly the disillusionment of these people that helped to create the dramatic collapse of faith in the period before the outbreak of the Second World War. More prudent supporters had never quite believed that the League could meet this supreme test. They were aware rather of the great value the League might have in various realms where issues were not carried to this extreme.

No one could doubt that the institution of something like the League of Nations (or the establishment of UNO at a later period) is a matter of great importance, partly because it brings the nations to a meeting place, and, still more, because it sets their representatives working together in cooperative tasks. Such an organisation, where it works in conformity with the spirit that engendered it, may be said to be knitting the nations together, or slowly marshalling them into what might be called an international order.

After 1919, however, there was a tendency to imagine that by the rather facile way of creating a new institution one had installed an international system or order ready-made – a thing which would not have been true, as yet, even if every participating government had joined it in the right spirit (which the idealists themselves were ready to recognise not to have been the case). To have a supra-national system which itself gives some support to the cause of order and morality, and provides a genuine sense of common life – a feeling that, in spite of quarrels, all are members of the same club – to have, above all, that finest and most genuine kind of system which generates the 'imponderables' (as a club or a college might do) multiplying the sentimental bonds between nations and turning controversy itself into more reasonable channels – all this is a matter of time and slow growth, and it needs considerable intervals of peace and normal, settled relationships. It needs countries which in spite of quarrels and discontents have come to see that they have a lot of good things to preserve – countries which, even if they have been revolutionised, have settled into the feeling that they have achieved a *status quo* so satisfactory that it must be defended.

In reality, the period which saw the rise of Stalin, Mussolini and Hitler must be reckoned as one of the most revolutionary, and one of the grimmest, in modern history. And even though eastern Europe, precisely because it had hitherto been so backward, might stand in certain respects as an exception, the future historian will surely decide that the total result of World War I (including the rise of dictatorships and the temporary barbarizing of Germany) was a blow to European civilisation and to the practical effectiveness of 'values' in the international sphere. Today one can be clear that it was the first stage in the downfall of the cultural leadership hitherto held by the Western half of the continent. Even among well-meaning men, as things developed, there was moral anarchy, leading to moral hysteria – a lack of any framework of reference (the loss indeed of just that mere minimum which the older 'states-system' had supplied). The well-intentioned were divided against one another and even against themselves – some moved by bad conscience about the past treatment of Germany, some blaming their own country for the failure of negotiations with Hitler or negotiations for disarmament, some calling for action against Hitler while opposing rearmament or saying that their government was not to be trusted with armaments, some asserting that the League of Nations had been too much the instrument of the victor-powers, or fearing that if war were undertaken too soon or continued too long, it might be condemned abroad as the attempt of the victor-powers to perpetuate a privileged position of which neutrals had already sometimes complained. In reality a carelessness about the power situation, a too moralistic attitude to international affairs, and a good deal of wishful thinking created situations which gave wicked men greater rope than perhaps ever before. The one thing which demonstrably was not achieved in the period between 1919 and 1939 was the thing that had been regarded as so easy – the provision of a guarantee against the rise of an aggressive power. And all the luxury of condemning the wicked does not compensate for the unwisdom of flouting what had hitherto been learned about the way of preventing their coming to the top. In fact there emerged – each initially on his own account – a bunch of three formidable aggressors, Italy, Germany, and Japan, all determined to exploit what had indeed become a case of international anarchy. Morality and good intentions had existed but had failed to secure a purchase on the international realm.

III

By 1945 the states of Western Europe had lost their leading position, and for them, as for the world at large, no opening was left for the self-hypnotisation of 1919. Those who held their breath in the subsequent period, as they watched the extension of Russian dominion in central and eastern Europe, and recognised the possibility that this could make itself permanent – those who wondered for some time whether the United States would ever awaken to the danger which still further Russian expansion might bring to America as well as to Europe – could now make no mistake about the factor of power in politics and the nature of the problem that is presented to the moralist. UNO may be carrying the world towards a future system, and its resolutions are not necessarily useless even when they are incapable of execution. But UNO does not provide at the relevant level the kind of order which enables men to steal a march on power; and, still, in 1970, world-opinion is unable to bring Israel and the Arab states even into negotiation with one another. The main object is obstructed if UNO comes to be used by particular interests or becomes the chief arena for the conflict between these, so that it does not really engender the feeling of belonging to the same club. And all difficulties are increased if its members have an overriding loyalty to a sub-system – the clique of former European Great Powers, the Afro-Asian group, the countries behind the Iron Curtain, for example.

If some of our predecessors thought that an international system had been impossible when the world was divided by a fundamental conflict of religions, we ourselves do not know as yet whether a genuine international order is going to prove feasible in a world that is divided ideologically. Here in particular, the danger comes when the greater loyalty goes to the sub-system – Protestantism or Catholicism in one age, Communism or Western Democracy in another. Since the Second World War we have had to ask ourselves sometimes whether the ultimate conflict does not even lie between conceptions of a supra-national order. In such a case both parties might be most ready to take part in UNO for the time being, but with the *arrière-pensée* that they would use it for their own purposes. The one might be seeking a system not unlike the order of things that existed in medieval Christendom. The other might be seeking to establish something more like the European 'states-system' of more recent centuries. If a struggle between two such patterns of international order lies concealed beneath all the amenities and all the quarrels of the present day international world, the one party could easily regard as 'moral' what the other party sees as 'immoral'. And this would be likely to apply particularly to problems connected with the rights of smaller states.

The autonomy of the smaller states of the world and the self-determination of peoples are peculiarly relevant to the problem of morality in politics, since they involve some sort of bridling of sheer power. And it is the reframing of 'self-determination' into a principle of 'anti-colonialism' which has characterised the latest period and provided the main example of ethical advance. In its new form the doctrine has found almost a mystique and has drawn to itself a greater concentration of moral force, acquiring a power which was hardly possessed by the doctrine of self-determination itself. This may be partly because the case inspires a special degree of sympathy and partly because it provokes a particular resentment. Partly also it might illustrate the point that a negative principle can be more powerful than a positive one, people always agreeing more easily about the thing they are wanting to attack. The question of Goa, and perhaps the attitude of some people to the problem of Gibraltar – though these present local complications – might suggest that where 'anti-colonialism' and a principle of local 'self-determination' clash, it is the former that commands the more effective global support.

The map of forces in 1919 was favourable to 'self-determination' in Europe, and in more recent times 'anti-colonialism' has owed a great deal to the redistribution of power at the end of the Second World War. The United States took up the cause, and possibly we might regard this as their great repentance, for, in a way, they are the arch-colonialists, the colonialists *par excellence*. The government of George III tried to save the Red Indians from the aggressiveness of these people. Indeed, if our brothers across the Atlantic escaped being culprits themselves in the 20th century this was because of the virtual disappearance of the population that held North America before the white man arrived. The emergence of Russia as a great leader of world-opinion was bound to imply (whether that country's statesmen took any action or not) that a powerful magnet was exerting a pull on the undeveloped peoples. If one reckons not only the two super-powers but also the uncommitted countries of the world, and if one adds the determination of subject-peoples to win their freedom, it would appear that world-opinion on this whole issue had become a political force. At the same time, among the possessors of colonies themselves there were springs of idealism, touches of conscience, together with some doubt as to whether their own populations would tolerate the extreme measures that repression would require. Moral ideas and political considerations seemed to achieve a sort of interpenetration by means of which they were able to reinforce one another. The Suez crisis of 1956 and the achievement of General de Gaulle in regard to Algeria represented no doubt particularly dramatic moments in the story. The final result must stand as a beautiful victory for a point of principle, even if it is true that the objective might have been reached in a better way – even, indeed, if it is true that a great price may still have to be paid for the freedom that has been acquired.

One paradox may emerge and its significance might depend on a finer knowledge than we possess of the deeper forces now at work in the world. If we look at, say, the fate of Poland and Hungary since the last war, or if we recall how the United States has extended its power, we may wonder whether Imperialism may not be finding new disguises, whether colonialism itself might not return in subtler shapes. Supposing such a thing were to happen it is not clear that the future historian would not be able to show that this grim line of development was really inherent in the general tendencies of our time. Perhaps we today would take the line that the movement – at any rate the surface movement – is in the other direction; for in the world at large there has been a remarkable expansion of freedom. But if something cataclysmic were to call for a right-about turn, one must wonder whether the railway lines are not already set for a possible movement in that direction too. When we think of Czechoslovakia in 1968 we may doubt the force of moral ideas or world opinion, and ask: 'Where was Man's guardian angel then?'

One thing seems clear, however. The consequences of the success of anti-colonialism – the consequences as distinct from the causes – have a real significance and seem to throw light on the processes of moral–political change. It is not merely that the old opponents of the movement have been gathered into its sweep, so that there is no question of directly reversing it; but the new situation has provided a fresh platform from which we survey the world – a fresh starting-point in particular for the thinking of a younger generation. Seeing the peoples of Africa as not merely possessing the right to self-government, but actually enjoying it, is a thing that does not quieten the conscience – it brings a higher sense of the wrongs that were being committed so recently. The mind becomes so re-shaped that it ceases to have any understanding – any respect – for its predecessors or for the past that is immediately behind it. Intelligent people emerge who find it impossible to believe that men ever did go out to the British colonies with fine intentions or any serious sense of mission.

There was born into the undergraduates of the last decade or so a greater compassion – at any rate a wider compassion in respect of the world overseas – than was apparent in similar circles fifty years ago. It may even be accompanied by a failure to realise that white men nearer home may call for some sympathetic understanding too. It helps to explain why, through practically a decade down to the tumult over Vietnam, 'Africa' was the magic word in so many of our universities. And now, more vividly than ever before, it is felt that, if there is hunger at the other end of the world, it is our own business to put a stop to it – if there is hunger at the other end of the world, we have a responsibility. Even if the young might be dismissed as Hippies, some of us who are older have long been feeling that the movement has been making similar advances inside ourselves. In this one singular sense we might wonder whether the state itself is not withering away – everything submerged in a feeling for human beings as such – international relations becoming much more than transactions between governments.

Sir Lewis Namier once mentioned the Jews and the British as the nations particularly guilty of legalism. It is not clear that the charge would be a valid one if directed against British diplomacy in general; but it may have had a certain applicability in the case of British diplomacy – and British attitudes – with respect to overseas possessions. One can feel virtuous when one is insisting on the sanctity of treaties or of constitutional enactments; but there could be no hope for international affairs if civilised nations were unable to do better than this, or if they used their *liberum veto* to prevent the amendment of treaties, the reform of the *status quo*. As far back as the 18th century one can find objections to the kind of treaty which uses the temporary defeat, the momentary incapacity, of even a European enemy to put that enemy under a permanent disability in a peace-treaty. In any case, the just man who honestly feels himself just may often be an obstruction to a good cause; and Justice itself can be an insufficient concept, for, mixed with a little legalism, it can be cover for a Shylock – for a nation that is insisting on egotistical rights. At least the supremely important, supremely interesting thing is the moment when the concept of Justice is itself revised, as, for example, when, in domestic affairs, it becomes disengaged from something like a deification of property. It means that love or charity must be above Justice and that diplomacy is in a position to go one better than law. And this may explain why in successive centuries of European history one meets the reiterated precept that foreign affairs ought not to be entrusted to lawyers. It may mean also that at any given moment, one ought to try to clear one's mind of concepts or slogans and simply ask the question: 'How can this conjuncture be most effectively used for the good of mankind?'

This whole development then – so profound in its ethical implications – shows how the world can progress in a period of general peace, even if under the shadow of the nuclear weapon. It gives a hint of the great things that often happen in periods of relative stability, and in spite of terrible counter-phenomena that threaten civilisation with cataclysm.

The smaller states of the world owe their freedom in the last resort to the existence of a global balance of power, though it may be essentially an equilibrium merely between two colossal giants. The nuclear weapon might well have rendered the concept less relevant nowadays, but this idea of the balance returned into currency after 1945, and it transpired that the former hostility to it had disappeared. An equilibrium established between two super-states is the most intense and most precarious form that the balance of power can assume. Irrespective of this consideration, the former apostles of such a system came to attach some importance to the establishment of subordinate regional balances. They would have regarded the freedom of European states as more secure today if there had also been a balance within the compass of the continent itself. A serious change in the distribution of forces might expose, in fact, latent tendencies towards the creation of satellites. And here the morality of the West might well conflict with the morality of the East.

The case of Czechoslovakia in 1968 reminds us of the importance of the principle of non-interference in the affairs of other states; and it shows how the cause of liberty can be served by this principle. It might be a mistake for well meaning states to surrender such a principle, save in an extreme case where a civilised state has gone barbaric; for we cannot hasten arbitrarily the political development of peoples. Yet sometimes the modern world is in a mood to dislike a principle that limits its desire to spread the good; and those who believe that they are the missionaries of a superior order of society may always hold that something like the Russian intervention in Hungary in 1956 or Czechoslovakia in 1968 is the higher morality, the higher service to mankind. The ultimate paradox arises (and the very efficiencies of the tyrant might be calculated to bring it about) when the subjected state comes to be glad that it has been disciplined into the required orthodoxy, just as Alsace, which had insisted that it was German, became reconciled to being French at the time of the Revolution. Such a retrospective vindication of a forced conversion is a subtle challenge to the principle that the Poles in 1945, the Hungarians in 1956 or the Czechs in 1968 have the prerogative of immediate self-determination. It could even happen that the outside world, seeing the subjugated country running to routine for decades, might cease to believe that it was unreconciled to its subjection. Hence the profound meaning of the final cry of the Czechs in 1968 – the prayer that the West should not forget them, should not imagine that the ending of protest implied their real acquiescence in unfreedom. It would be prudent to assume that the freedom of small states rests even today on the balancing of power against power. And the world in general does in fact rely on check and counter-check rather than upon UNO for the continuance of such liberty as exists. It is not even clear what would happen to the world of small states if the massive watch dogs were called off – that is to say, if both Russia and the United States were eliminated as leading powers and invisible constrictions were removed. The result might well be anarchy and the loss of liberty. The best supports for a purely moral check on aberrations would be time, protracted habits of peace, the slow growth of reasonableness, and conditions of normality already experienced, enjoyed and valued.

In the latest period of all there seems to be a wider moral change which cuts across the accumulated experience of the human race, and contravening the teaching of centuries concerning the decline of states and the downfall of systems of freedom. Such a tendency would not necessarily be disastrous; but it implies an overthrow of safeguards which both pagan and religious teachers have hitherto thought essential; and therefore one would wish to be sure that the matter had been deeply considered. It is right that mankind should move forward to new adventures in living; but civilisation requires a peace and internal order which depend on vast efforts of human contrivance carried on over long periods of time. Simply to set the world loose on a flood of uncontrollable, unforeseeable, change might bring irreversible catastrophe to a democracy.

It even remains to be discovered whether the combination of the welfare state, the permissive society, a body politic dominated by materialistic purposes and governments preoccupied with the need of catching (or the fear of losing) votes can generate the moral qualities necessary for the survival of states. Perhaps it still has to be discovered whether (short of the installment of appalling tyrannies) technological society is not itself a more vulnerable and altogether more sensitive affair than has hitherto been imagined – open to a vast, further extension of sabotage and helpless against numerous minority groups, each of which, realising its indispensability to some part of the system discovers at last the formidable character of the weapons it possesses. Such a society may require for its continuance not merely passive acceptance (like the older ones) but the positive collaboration of its people – a thing not always easy to maintain. Alternatively, we might even be moving into a world in which the hardest heads, the biggest organised interests and the managerial minds will come to see communism as the only way of integrating society again and regenerating authority. Over and above all this, a world that cannot cope with the elemental things that are so relevant to the existence of an order – the infringement of diplomatic immunities, the virtual taking of international hostages, high-jackings in the air (things which the ingenuity of man will be able to multiply) is possibly on a dangerously inclined plane.

Somewhat more closely connected with the transition from one generation to another, however, is the recent emergence of heresies, fallacies, types of opinion and patterns of conduct relating to international politics which carry too many reminiscences of the inter-war period, as well as too many similarities with what has been predicted in regard to the dangers and weaknesses of democracies in the field of foreign policy. Combined with the shortness of the popular memory in this field, these might be in danger of generating problems within the society that is affected by them. One is justified in inferring that modern education needs not merely superficial teaching about current affairs but a disciplined study of the deeper nature of international politics in general.

We have learned in these latest years how easily the cry of 'American Imperialism' could be raised against a country which (whether it deserves the charge to any degree or not) was certainly less willing than any of its predecessors to assume the lordliness or the responsibilities of a world power. We have heard of American 'aggression' and a new generation often does not know (and does not credit the fact when informed) that Western Europe once wondered whether the United States could ever be awakened to the danger from Russia. It is true that, long after that time, even supporters of America might have misgivings concerning a certain self-righteousness associated with her tradition and connected perhaps with inexperienced democracies: she was to have too clear a sense of representing right *versus* wrong. The provision of a check on Russian power seemed to involve the Americans in even too religious a hatred of communism, which in spite of atrocities and recent aggressions had conferred a number of benefits upon the world. Indeed, something can always be said on behalf of 'guided democracies' where a people are undeveloped and something can be said on behalf of Communism, not for Europe perhaps but for Asia – that is to say, where agrarian revolution of a drastic kind might be the supreme requirement. There arose among us also the misgiving that the United States, like other democracies, having closed her eyes to the problem of power and gone to excess in its depreciation of power politics, might now be running to the other extreme: too confident that all was well since it was righteousness that was being armed – too ready even to exaggerate what could be achieved by power or to imagine that, in case of difficulty, all that was needed was to throw in an additional load from her vast accumulation of forces. From as early as 1962 one could be sad about the Vietnamese adventure and feel that perhaps the affair was regarded as too easy, or that there was an idea of getting around all difficulties by pouring in more power, while in fact (in spite of any pretence that America was not really involved) it was difficult to envisage any likely disentanglement from the affair in the event of misfortune, considering the way things generally happen in the world. It was possible to imagine that from the first, strategic objectives or strategic fears were mixed with more generous motives in the case of America's original involvement in Indochina; but, even so, one would not have dreamed that the enterprise itself would quite be charged to American imperialism, American aggression or American militarism. Indeed, it was possible to deplore the whole idea while still believing that it had been undertaken with a heavy heart.

In the realm of international politics, below the crimes of particular people, there is a fundamental human predicament which conditions both the men and their conduct, conditions even the virtuous, helping to generate their fears and to shape their policies. It lies deeper than the conflict of ideologies and, for example, one must never imagine that all would have been well for the post-1945 world but for the existence of communism. For a century before 1914 – and long before there was reason to think of communism – there existed recurrent fears of the Russian bogey, a terror of what might happen when that colossal empire properly organised itself. And this was postulated upon a Russia that would be Tsarist and even Christian – it assumed only the ordinary workings of human nature. In four centuries of modern history the successive appearances of Spain, France, Germany and Russia as the aggressive states, 'the menace to the continent', have coincided closely with the moment when these countries respectively acquired the capacity to take up the role and so found themselves open to temptation. Even if they had committed no offence, their power would have provoked fears or jealousies and therefore precautionary measures on the part of their neighbours – these things in turn producing counter fears and counter measures which to the rest of the world would be indistinguishable from aggression. One great power may confront another without either of them (or indeed their historians for a considerable time) being able to measure or understand or credit the other's fear. The human predicament in the twentieth century would not be greatly or enduringly altered if Russia and even communism were to disappear entirely from the map. The forces would certainly reshuffle themselves, and the pattern would change, but the fundamental problem would not be solved. Even a *détente*, though desirable and not impossible, would not eliminate the predicament; and if one party were to be misled by it – hoaxed into taking off the safeguards and abandoning its vigilance while the other were unremittingly alert – the result would be an irreversible misfortune. There are too many people who in politics can see only one thing at a time – a single right set against a single wrong, for example – and they evade the real problem by resorting to an oversimplified moralism. One of the great tests of statesmanship is the catholicity with which it can allow for all the factors in a complicated case. If the government of the United States has failed to avoid some of the pitfalls, the fault may be due to an oversimplified view of righteousness rather than any lust for power as this is ordinarily understood. And this may even be the defect also of its opponents at home – for liberalism may miss opportunities for originality and depth if it evades the intellectual challenges. At least in democracy, liberalism can only advance by promoting the slow growth of reasonableness among men.

When everything has been said, the world always underestimates the operation of fear in international politics – perhaps also the secret movements and machinations of fear in life as a whole. Sir Edward Grey was not lacking in vigilance, but on one or two occasions he made wise and realistic remarks about the dangers of over suspiciousness in diplomacy. History, as we know it, would be unthinkable if it were not possible for men to steal a march on power – definitely steal a march on it and not merely supersede it with greater pride. One could conceive the next step in progress to be one which would involve a crucial act of faith in human nature, in spite of all that has been said about the limitations of this latter. It would be a thing not without risk – a risk which itself would have to be very carefully calculated; and it is a measure that could be achieved only by a state that would be acting from a position of power. It is not the young men or the academic people, or the wishful thinkers who would ever be able to measure the chances of it – and those who merely consider the struggle as a straight war of right against wrong would never entertain the issue at all. For such an object it would almost be better to confide all that one values to the good intentions of a Bismarck, in spite of all his offences – confide it to a Bismarck at his best. There could be no higher act of statesmanship than this and no act that could ever require a more creative statesmanship.

If there exists any hope in this direction there lies, possibly closer at hand, a more solemn danger which involves both morality and morale and might have serious results. It comes from the fact that Western Europe pins its faith on a policy which involves the threat to be the first to resort to nuclear weapons, in spite of serious warnings that this would lead to a great escalation. Such a policy might have been understandable at a quasi revolutionary moment when a relentless aggressor was already on the move and a direct attack might be felt to be imminent, so that desperate measures were excusable for the time being. It is out of context in a world where conflict, if it arises, is likely to spring from complicated situations – indeed in a world where one is hoping for a *détente*. It loses sight of the fact that in nine cases out of ten a movement of Russian conventional forces could arise (and perhaps it could always be made to arise) from a crisis which was so morally mixed and muddy that the resort to nuclear weapons would be unthinkable. Any gamble made by Russia, if she called our bluff in such an event would rest on an assessment before her eyes, while the West is making a permanent gamble on some future situation that is always unpredictable. The threat is not a whit less irresponsible if there is an *arrière-pensée* of not keeping it. The word 'deterrent' now becomes the greatest hoax of all where it carries the implication that the thing can operate of itself so that one can mentally extract from the discussion the very weapon upon which its working is supposed to depend. The whole idea points ultimately to a self-righteousness which has become institutionalised and organised and which represents one of the biggest breaches that our world is making with Christianity. The West would receive a terrible punishment if such a policy came to failure; and, for a long time, a Christianity so closely associated with the West would go through the darkest of tunnels.

Note

1. Original in BC File 110 (eds).

22
The Balance of Power

The idea of the balance of power is associated with the modern history of our part of the world, and envisages the political units of the Continent as forming what used to be called 'the European states-system'. On this theory, the whole order in Europe was a kind of terrestrial counterpart of the Newtonian system of astronomy. All the various bodies, the greater and the lesser powers, were poised against one another, each exercising a kind of gravitational pull on all the rest – and the pull of each would be proportionate to its mass, though its effect would be greatly reduced as it acted at a greater distance. When one of these bodies increased its mass, therefore – when, for some reason, France for example had an undue accession of strength – the rest could recover an equilibrium only by regrouping themselves, like sets of ballet dancers, making a necessary rectification in the distances, and producing new combinations. Otherwise, the overgrown power would swallow up the little ones near at hand, and become greater still – just as the moon would fall into the earth if there were no counteracting forces to offset the effect of gravity.

David Hume began his essay on 'The Balance of Power' with an imposing list of classical examples; and he has sometimes been taken as the authority for the assertion that the principle was familiar to the ancient world. He recognises the fact that his view was seriously challenged in his own day, however; and it is clear to me that the examples often quoted to support his thesis do not quite meet the case. This is true of the passage of Polybius which Hume instances, and which, because it so excellently formulates one of the elements of the situation, is among the most famous historical texts on this subject. In it, Polybius applauds the ruler of Syracuse for helping the Carthaginians to resist the rise of Rome; and he puts forward the maxim: 'It is never right to help a power to acquire a predominance that will render it irresistible.[1] I suppose that, if we today read our modern ideas into Polybius, we can say that in this thesis he is demanding, to say the least, that there shall be a distribution of power. But a passage of this kind would not (and historically it *did* not) suffice to communicate the idea of the balance of power to people unacquainted with the idea already.

In discussing the states of the ancient world, Hume makes reference to their way of 'shifting sides' sometimes in periods of conflict; but he knows that states change sides out of feelings of jealousy or a sense of emergency, and that it is merely our hind-thought which so easily turns these policies into a doctrine of the balance of power. When people denied that the ancient world was aware of the doctrine, Hume replied that the Romans had never known of it, but the Greeks had possessed it, and the moderns had studied the history of the former while neglecting the latter. Rome, he says, had been helped in her rise to supremacy by the fact that the states of the Hellenistic World were unaware of the principle of the balance of power. What seems to surprise him is his discovery that the ancient writers had never called attention to this defect. From all this I should infer that the idea of the balance of power not only did not exist in the ancient world, but did not take its rise even from the modern study of ancient history. More than most of our basic political formulas, this one seems to come from the modern world's reflections on its own experience.

It is generally understood that the states of Renaissance Italy, like those of ancient Greece, formed a neat closed area in which the principle might be expected to develop. It provided an arena of limited size – something like a miniature states-system – a field of interacting forces such as can be envisaged for the most part in isolation. There is no doubt that, within such an arena, states will want to support the power of which they are less afraid against the power of which they are more afraid; so that interesting combinations are likely to be produced. At times, the governments in Renaissance Italy seemed to be playing with weights and counter-weights, and calculating, among their neighbours, the tipping of the scales. After the long struggle between Florence and Milan, it came to be seen (Hans Baron tells us) that the Venetians had been standing between the two belligerents, trying to secure first that the conflict should go on long enough to weaken both, and secondly that it should leave the two fairly equal in strength. We even meet with the word 'balance' on occasion, though we may wonder about the picture that was made of it, and the degree to which the notion was followed up. Since an idea can sometimes be illuminated by studying its negative, it may be useful to illustrate the effects of the absence of the doctrine of the balance of power. These can be seen even in the cases of Machiavelli and Guicciardini, who represent the crude beginnings of what might be called the science of international politics.

Machiavelli was shocked by the weakness of Italy's international position in his day. The tragedy of foreign invasion was one of his excuses for his enquiries into the science of politics. His own experience of public life is chiefly known to us by the records of his diplomatic missions. He devoted a great part of his study and reflection to problems connected with war. One would have expected his teachings about policy to be impressive, therefore, in the field of foreign affairs. But, though he is rightly famous for having established the beginnings of modern military science, his work seems to me to have been particularly weak on the diplomatic side. He always insisted on the importance of following the example of classical antiquity. Yet, for him as for other people, the ancient world proved a better quarry for military teaching than for maxims of international politics. He imagined that in his own day a kind of dictator might be needed to save Italy from the foreign invader; and I always feel that his friend Guicciardini had more diplomatic sense, and a better eye for the international scene. Guicciardini said that the way to save Italy from invasion was to take care that the powerful nation-states around her should be kept too busily occupied with one another elsewhere.

Machiavelli is weak in respect of the principle of the balance of power; and his weakness is all the more remarkable, his deficiency all the more clearly exposed, because his preoccupations ought to have drawn him to it. He had an unusual interest, even an obsessive interest, in one of the associated problems (an issue similar to that which Polybius was discussing in the passage to which I have referred), and he is responsible for awakening a wider interest in this particular problem – one which lasted throughout the 16th century. The point is worth examining, perhaps, because it illustrates the kind of thinking which men did before there was a current doctrine of the balance of power – a kind of thinking which was scarcely possible afterwards.

The problem and the controversy in question were ones which hung on the issue: what ought a state to do when its neighbours are at war, and one or other of these neighbours is going to increase its strength as a result of the war? It would hardly be possible in more recent times to approach this kind of theme without raising the whole issue of the balance of power. Neither Machiavelli nor his successors had any idea that the case could have any connection with a notion of the balance of power, and that the debate on the whole subject would ultimately be subsumed in discussions of the balance. Machiavelli, however, merely treats the topic as a question of the wisdom of remaining or not remaining neutral when one's neighbours are at war. In fact, in his most famous book he gives himself away in a still more significant manner. He deals with the matter under the curious chapter-heading: 'How a prince must act in order to gain reputation'.[2] Elsewhere, also, he tends repeatedly to treat this issue as one which primarily involves the mere question of prestige. On one occasion he says that you had better be sure to be on the winning side, though he knows that the victor may be ungrateful at the finish; but he holds (though it is unlike him to entertain such an illusion) that this will not usually be the case. What he is really sure about, really insistent about, is that both parties will hate or despise the neighbour who has merely remained neutral while they themselves have been at war with one another. (There is one occasion on which he advises an alliance with France, because France is likely to win, at any rate if she secures this alliance; though he adds that France, even if she wins, will be more tolerable as a victor-Power than the other party can be expected to be.)[3] He comes nearest to Polybius when he says in one place that 'a prince ought never to make common cause with one more powerful than himself with the object of injuring another, unless necessity forces him to it'. I am not sure that these various ideas are very acute and I am very doubtful that they can be squared with one another. On the kind of issue which seems to me to distinguish ancient from modern discussions of foreign policy, I should regard Machiavelli as standing with both feet in the ancient world. I think, therefore, that we ought not to attach too much significance to the fact that, at least on one occasion, Machiavelli did make reference to the existence of a balance in Italy. His younger contemporary, Guicciardini, has a similar interest in the question of the wisdom of remaining neutral when one's neighbour's are at war. He is more subtle than Machiavelli here, and as usual sees that there is a further range of considerations to be taken into account. He writes, for example:

> To remain neutral when others are at war is a good thing for a state that is powerful; for such a state need not be afraid, whichever side may win. It can preserve itself without difficulty and it is in a position to gain profit from the disorders afflicting the others.

Neutrality is dangerous, however, for a power that is weak, he says. It merely results in your finding yourself ultimately at somebody else's mercy. Neutrality is most dangerous of all if it is the result of irresolution, and this is what often happens in the case of republics, where opinion tends to be divided. In this case, even the belligerent who was wanting you to be neutral, would be unable to feel sure of you, so that both parties would be left dissatisfied.[4] If you make alliance with either of the belligerents, he says in one place, your fears are at least reduced to one; for all that you have to worry about is the possible victory of the other party. Neither Machiavelli nor Guicciardini seems to me to be very interesting or profound on this issue; and I have the feeling that there is something archaic in their handling of the problem, because they are so constricted in regard to time and in regard to place. They are thinking of an *ad hoc* decision at the outbreak of a war, without reference to diplomatic action in its continuity, and in its long-term aspects. At the same time they fail to consider the overall international scene, the general disposition of forces in the world.

Yet it was Guicciardini, who ultimately made the crucial advance and gave the first vivid picture of the balance of power. He did it in an account of the political situation in Italy before the death of Lorenzo de' Medici in 1492; and the passage occurs in the very first pages of his *History of Italy* – a part of the work which only recently has been discovered to belong to a late date, that is to say, the year 1537. According to Guicciardini, Lorenzo de' Medici, the ruler of Florence, decided that his city would be endangered if any one of the major powers of Italy succeeded in achieving further aggrandisement. He determined therefore that the affairs of Italy should be kept in balance – one side not more weighted than another; and for this purpose, he saw that it was necessary to have peace. The King of Naples also wanted peace, and wanted to prevent changes in the Italian peninsula. The ruler of Milan knew that he, too, had more to fear than to gain by any disturbance of the existing situation. These important Italian powers came together, therefore, and formed a combination which proved strong enough to discourage foreign invaders, strong enough also to check the Venetians, who were greatly feared at this time. And, though it is important to note that they were not sincere in their friendship with one another, their system did achieve the object they had in view. Indeed, almost all the minor principalities of Italy came in to swell the alliance, we are told; and this makes us raise our eyebrows a little, provoking the question whether it really did represent a 'balance' of power. In a profound sense, it was genuinely a balance that Guicciardini set out to describe, and the crucial evidence for this is a significant section of this descriptive piece. He wrote of these Powers:

> Since they were moved by mutual jealousy . . . they were unremitting in the watch that they kept on one another's movements, deranging one another's plans whenever they thought that a partner was going to increase his dominion or prestige. And all this did not make the peace any less stable, but rather made the powers more alert and more ready to bring about the immediate extinction of all those sparks that might start a fire. Everything was so disposed and counter-balanced that not only was there no fear of change in the present – it was difficult to imagine what counsels or dynasties or armaments could succeed in disturbing such a peaceful order in the future.[5]

The historical accuracy of this particular picture of late 15th-century Italy is not the issue that concerns us at the moment. In this – perhaps the most famous of all literary passages on the subject of the balance of power – we have the distinct impression of a system of forces which has been brought to an equilibrium. It is not a perfect picture but it is remarkable for a reason which Francis Bacon particularly noted – namely, the suggestion of nervous tension throughout the system – the various principalities not really trusting one another, but jealously watching one another's every move, diplomacy being unremittingly awake, and the whole still serving the purpose of peace. Guicciardini insists on the point: he says that Florence could not have kept the system in balance if she had not 'shown the greatest diligence in watching every turn of events, however small'. At a later date, Francis

Bacon particularly caught up this point. He talked of princes 'keeping due sentinel'.[6]

Guicciardini painted an interesting scene – a particular situation existing in Italy towards the end of the 15th century – but we must not imagine that he expounded a general theory of balance, or that the doctrine now became current in the world. In fact, from the time of the Renaissance, the reference to a 'balance' in international affairs generally shows that something was lacking – the real appreciation of the doctrine had just been missed or the formulation of it went wrong. At the time of the Renaissance the French diplomat and memoir-writer, Philip de Commynes, described France as confronted by England, England as set off against the Scots, Spain as being the opposite number of Portugal, and Bavaria as competing with Austria. He speaks of each state or dynasty as having its eye on its opposite number, but what we miss here is the notion of a general field of forces, the idea of what we call a states-system. He says somewhere that, against each government or princely house in Italy, God has set up a 'contrary' to keep it humble – the Venetians against the Florentines; in Naples the house of Aragon against the House of Anjou; in Milan the family of Orleans against the Visconti.[7] Henry VIII, when he met Francis I of France at the Field of Cloth of Gold in 1520, had a device which represented him as holding a balance; but he was far from suggesting that the European states should be in equipoise. His motto merely promised 'Cui adhaereo prae est' – the one that I join is the one which will turn the scales. Even at a later stage than this, I get the impression that it was possible for people to have a negative idea of the balance without giving much thought to what constitutes the actual equipoise. It was an example of what used to be called the 'neglected positives' – the opposites of words like 'uncouth' and 'unkempt' – you knew when a preponderating Power had created a situation of imbalance, but you had not examined the notion of equilibrium itself. I think that it would be agreed nowadays that historians were once too ready to read back the doctrine of the balance of power into Tudor England. In any case the theory of the balance of power is to a great degree unachieved so long as people are thinking of a pair of scales and merely using it as a figure of speech.

Towards the end of the 16th century, there was a famous scholar, Justus Lipsius, who had an extraordinary influence as a teacher of statesmanship, whether you measure the influence by the number of translations and editions of his works, or by the way his teaching was taken up in universities, or by the fame of the monarchs and statesmen who sought his advice and tried to enlist his services. It was he who did so much to redeem the reputation of Machiavelli by separating the scientific aspect of his teaching from the maxims that were so obviously unethical; and it was important that he – a man palpably on the side of morality – insisted on recognizing Machiavelli's doctrine about the importance of force in international politics as well as his idea of learning politics from antiquity. But this man knew nothing of the balance of power, so far as I can see. He talked the language of Machiavelli and merely put forward the doctrine that it is a mistake to remain neutral when other Powers are at war. Francis Bacon comes nearer to the modern idea; for, when he talks about Henry VIII, Francis I and Charles V, he says that, if ever any one of these rulers acquired a foot of earth, the other two would immediately set about to restore the balance.

It is my experience that if, as a general student of history, one collects references to the balance of power, or studies the ones that the experts have collected, they are comparatively few in the 16th century, but, after 1600, not only do they become more numerous – their meaning is less clouded by ambiguities. It is after about the middle of the 17th century, however, that the references begin to come in an amazing flood. And, so far as I can see, it is only at this point in the story, that the doctrine has its remarkable development.

There are references to the balance of power in the despatches of Mazarin in the 1640s, and in this period the idea was particularly associated with Venice – in both French literature and French diplomatic documents you find the Venetians described as the apostles of the doctrine.[8] But it is the course of events in the subsequent decades of Louis XIV's reign that gives the theory its extraordinary currency. Now that France had replaced Spain as the menace to Europe, it came to be seen that it had not been the peculiar wickedness of the Spaniards that had previously threatened the world – it was the disposition of forces that made the Spaniards the aggressors in one age and then the French in another age. The doctrine of balance was often deprecated in this new epoch by the French, for of course it became an important weapon against them as Louis XIV proceeded in his career of aggrandisement. And the significance of the whole theory was magnified because governments were paying attention to propaganda – Louis XIV's wars provoked in various countries great numbers of pamphlets and topical treatises. The War of the Spanish Succession was a remarkable example of the way in which the policies of European states came to be affected by the doctrine of balance. By this time the doctrine is repeatedly appearing in diplomatic despatches, state papers, treaties of alliance and treaties of peace.

Even in this new period, the idea was still rather rough at first; and some writers said that, between France and Spain, the Dutch represented the tongue of the balance – others said the English. It has been pointed out that this was a haphazard use of metaphor. Since the doctrine came to be directed against Louis XIV, official circles in France tended to dislike it, and support from that country came from critics of the government. In the ranks of these was the famous Fénelon who produces for the education of the Duke of Burgundy a work that showed a high degree of sophistication.[9] First he said that if once a Power was allowed to rise to a position of predominance, you would not be able to count on its good behaviour, however moderately it had hitherto behaved. Once it could act with impunity, it would not confine its ambitions within the accustomed channels. In fact, said Fénelon, you could not expect it to remain moderate for more than a single reign. Secondly, Fénelon pointed out that even a nation which at one time has been helping to check an aggressor will in the very course of those proceedings move imperceptibly into a career of conquest. In other words, it demands more and more securities against the old aggressor, and, suddenly, when you turn round, you see that it has been moving towards universal dominion itself. Thirdly, Fénelon is remarkable because he held that the balance of power should be regarded as an overruling law. The internal laws of a country – the law about the succession to the throne, for example – should give way to 'the right that so many nations had to security'.

Once things had reached this point and the idea had come to such conscious formulation, the balance of power not only became a diplomatic objective but was exalted into being the very highest of such objectives. The element of egotism in state policy was clearly recognised all the time; but in so far as you were not being egotistical, you were supposed now to aim at the preservation of the international order, the maintenance of the balance of power. The principle did in fact prescribe limits to egotism and ambition, and the check did actually operate. It operated because it involved a more enlightened view of your own interests – it was a case of limiting your short-term objects for the sake of your long-term advantage. I wonder whether, along with Bismarck, perhaps Frederick the Great is not a remarkable example of a conqueror who – unlike Napoleon – accepted a principle of self-limitation, seeing in the long run the importance of an international order, showing some solicitude for that order, and turning into something like a conservative statesman. And, of course, under this system you could be egotistical but at least even the egotist had to pretend to be following the principle of the balance of power, demonstrating that there, at any rate, was the accepted standard.

I do not think it can have been an accident that the doctrine of a European equilibrium became so fashionable, and proved so pregnant, just at the time when the world had become familiar with parallelograms of forces, and men were beginning to see the heavenly bodies beautifully equipoised. The expositions of the theory sometimes pressed mechanistic analogies and figures of speech in a remote way, and the sentences or the analyses come on occasion with a curiously Baconian ring. This was a general tendency, as the 18th century proceeded, and I recently read an unpublished essay which showed how important the idea of the balance was to Gibbon, whose particular predilection was the pendulum. Along with the idea of the balance of power, the notion of the balance of trade and the theory of equipoise in the British constitution seems to have been becoming fashionable in the 17th century. Francis Bacon has a rather different notion of the 'well-balancing' of trade, and returns at times to the idea of having the parts of society in equilibrium – the nobility in due proportion. Also he recognises that a change in trade-relations can alter the balance of power.[10]

It is easy to see why the 18th century attached so much importance to the doctrine of balance. It constantly remembered the aggressions of Louis XIV as things that must never be allowed to happen again. It rejoiced that now there was no single Power which could be described as having the predominance, lording it over the continent. It remembered in the same way that in the 17th century a vast Swedish Empire had encircled the whole of the Baltic, and that, as this collapsed, the rise of Russia might bring another Power into the dominant position there. But for the time being all was well – the 18th century prided itself on the fact that the Baltic was now divided between five different states, no single one of which could be regarded as having the predominance. In the background of 18th-century thought there was the repeated remembrance of a past, still fairly recent, but darker than anything else—the cruel Wars of Religion, when one party fought to make all Europe Catholic, and the other to make it Protestant, without any regard for the balance of power. The 18th century did not set its heart on either a Catholic order in Europe or a Protestant order, but on an international system which was to be defended for its own sake; a new kind of order because it comprised both Catholics and Protestants, just as it comprised both monarchies and republics.

But, more remotely still, the 18th century looked back to the Roman Empire as a thing that must never be allowed to happen again. They realised, what the twentieth century forgot sometimes, that there are only two alternatives: either a distribution of power to produce equilibrium or surrender to a single universal empire like that of ancient Rome. And this development in their theory became supremely relevant when Napoleon overthrew the balance and seemed to be creating a new Roman Empire. In other words, the theorists of the balance of power found it necessary to vindicate the idea of a Europe divided into separate states of only moderate size—a map of the continent which had the appearance of a patch-work quilt. Against the notion of a uniform Empire with a uniform culture, they promoted the idea of a civilization fundamentally one but broken into panes of many-coloured glass—achieving greater richness through the variety of its local manifestations. In general, and in a fundamental sense, it was their case that the distribution of power was the important thing, but that, apart from this, small states in themselves had an intrinsic value. The balance of power was a system—the only system—which, in a world so much at the mercy of force, made it possible for a considerable number of small states to remain in existence at all. Granted such a European order, it was not at all necessary – nor was it possible – that the various states should be equal or even roughly equal in power. In fact one writer, Gentz, asserted that, if Europe were divided up into states, that were absolutely equal squares, it would actually be harder, not easier, to establish between them anything like a balance of power.[11] Gentz pointed out that, even when one state increased its territory, it was by no means always necessary that the other should do so. The rectification of the balance might normally be brought about by a reshuffling of alliances. Furthermore, the balance not only guaranteed the existence of small states, but assured them of certain autonomy, a power of independent action. It guaranteed their independence at its most crucial point, namely, in the realm of foreign policy. In other words, it made possible the existence of small states that should not merely be satellites. Richelieu had once pointed out that in the existing system, small states had greater freedom of action than larger states – in the case of an alliance it was the larger rather than the smaller power that was likely to be abandoned by its partner.[12] And in fact the system of the balance of power even depended on the free action of the smaller states which, as an ally of theirs became too powerful, could shift their allegiance and throw their weight on the opposite side.

From all this you will see that the 18th century conceived of the balance of power in an imposing way and elaborated it into a whole theory of international relations. But there would be differences about the details of the system. There was a school of writers in Hanover who tended to regard Great Britain as the standing supporter and guardian of the balance of power. They took the line that a maritime state is never a danger to liberty. The French had a different point of view, and they sometimes argued that England was an offensive intruder – she really belonged outside the alliance-system. England, they said, talked about the balance of power in Europe but showed no respect for the principle in the rest of the world. The French complained in a similar way against any government that allied itself with Russia; and they affected to think of Russia as rather an Asiatic Power, that ought to be excluded by general assent. About Turkey there was bound to be much difference of opinion – in the 1790s the question was raised in the British parliament whether she was within the system or outside it. In some respects she had to be reckoned as being inside the balance-system, because she had a traditional alliance with France.

The mechanics of the European order – the design of the whole machine – was sometimes a complex affair; and some writers describe wheels within wheels. It was possible to conceive of many local systems of the balance of power, and then to comprehend all these in a wider system of European balance. One writer sees first an equilibrium between the south-western states, then another among the north-eastern states, and then a third within Germany, and finally a comprehensive one, embracing all these and spanning the whole continent. After 1763, when both Russia and Prussia had emerged to greatness, you had a curious triangle of forces in Eastern Europe – Austria, Russia and Prussia all poised against one another, all crouching like tigers ready for a spring. If one gained an advantage the other two would draw together to redress the balance and secure compensation – there was constant switching and interchange among the three ballet dancers – it is the most intensive application of the principle of the balance of power I have ever seen in history. But if those *three* Powers ever agreed on a policy, the Western states – England and France, for example – could never stop them in Eastern Europe. And that is how the Partition of Poland was able to take place – there might be a balance in Eastern Europe, but there was a defect in the overall European balance-system. On the other hand, for a long time the balance of power enabled the Turkish Empire to survive after it had become too weak to defend itself. Here the West was able to counter-balance the East – England and France were able to check Russia – partly no doubt through the existence of sea-communications.

The principle of the balance of power apparently tended to the preservation of the *status quo*, putting a brake on territorial changes. In one of his Political Testaments, Frederick the Great marked out a piece of territory for future acquisition but gave warning that such an attempt at aggrandisement was not feasible at the time.[13] However completely you took the world by surprise, he said, the other states would intervene, and sides would be too equally matched. The struggle would be a severe one, and, at the finish, the game was not worth the candle. Within a few years of this, Edmund Burke was writing in the *Annual Register*:

> The balance of power, the pride of modern policy, and originally invented to preserve the general peace as well as freedom of Europe, has only preserved its liberty. It has been the original of innumerable and fruitless wars. That political torture by which powers are to be enlarged or abridged, according to a standard, perhaps not very accurately imagined, ever has been, and it is to be feared will always continue a cause of infinite contention and bloodshed. The foreign ambassadors constantly residing in all courts, the negotiations incessantly carrying on, spread both confederacies and quarrels so wide, that whenever hostilities commence, the theatre of war is always of a prodigious extent. All parties in those diffusive operations have of necessity their strong and weak sides. What they gain in one part is lost in another; and in conclusion, their affairs become so balanced, that all the powers concerned are certain to lose a great deal; the most fortunate acquire little; and what they do acquire is never in any reasonable proportion to charge and loss... (In) modern treaties of peace... none can properly be called conquerors or conquered.[14]

Gibbon says much the same thing but puts a better face on the matter. He wrote: 'In war, the European forces are exercised by temperate and indecisive contests.'[15] But it was sometimes recognised by the theorists that the balance of power, precisely because it vindicated small states, operated to preserve freedom rather than peace. It corresponded to one of the peculiarities of the European order – an order in which freedom was more important than tranquility. In other words, you have to choose between the principle of universal empire and the principle of the distribution of power.

An article in the *Edinburgh Review* in 1802 declared that the idea of the balance of power was the result of the progress of science and the peculiar circumstances of the modern world. It was as unknown to the ancient Greeks and Romans as the teaching of Kepler and Newton. The article emphasised the elasticity which the system required – constant vigilance, careful observation of little changes that take place, perpetual interest in what was happening at the other end of the map. It took the line that you should check the potential aggressor before he really emerged; in fact what you ought to attack are the conditions that make for aggression – what you ought to prevent is the excessive aggrandisement of any Power. It is too late to take action afterwards; it is a mistake to wait till the aggressor has actually shown his hand. The *Edinburgh Review* takes the line that the various Powers have the right to control one another – to prevent one of their number from becoming dangerous. I have come across the assertion that, if a state is becoming too powerful, it is legitimate to set out to weaken it even before it has made any aggressive move; though there were differences of opinion on this question. Francis Bacon held that if a Power had made itself formidable, your fear of what it might do was a sufficient justification for going to war. Grotius, however, had insisted that war was wrong unless the overpowerful state was itself about to attack. One could add that the whole system required what might be called an intelligent appreciation of the international order as such. Everything was liable to be ruined if Englishmen hated the French too much and imagined that nobody but a Frenchman could ever be an aggressor. It was generally our fault that we went on fighting the old aggressor too long and thereby made it more easy for a new aggressor to emerge. It could be equally dangerous if the English as in 1814–15 loved the Prussians too much, and consented to excessive aggrandisement for them, on the assumption that they would always remain friends.

In fact, rigidity was always a serious danger and the system was under constant threat through the existence of deeply rooted prejudices. From the time of the Treaty of Utrecht, there was a tendency to regard the system as stable for the time being; and there existed a presumption in favour of keeping things as they were. There was even a version of the doctrine that tended to rigidity – a way of speaking about the balance of power as though it were the 'constitution' of Europe – a constitution partly written (in treaties, for example) but partly unwritten (like the balanced constitution of Great Britain). Gentz, who defended the whole system in the time of Napoleon, wrote:

> What is generally known as the balance of power is that constitution of neighbouring and more or less connected states, by means of which no one of them can damage another in its independence or essential rights without being restricted somewhere, and therefore endangering itself.[16]

In reality, statesmen and writers would often tend to identify the balance of power with the existing map of European forces, disliking any change that would entail a distribution of current alliance-arrangements. Alternatively, various states would set out to preserve what they regarded as their own system of the balance of power. States might continue alliances because they had become sentimental, or alliances might remain too fixed because of marriage-connections or commercial interest. The French, when in the 18th century they talked of maintaining the balance, often meant the preservation of the existing alliance-systems instead of rapid adjustments to a change in the distribution of power. In fact, by their attachment to Sweden, Poland and Turkey in this period, they may have been too lacking in flexibility, while Russia was coming to the front. On the other hand, if interests of state placed you in one alliance-system, but you had a standing marriage-connection or a regular commercial link with a member of the other group, this might help to form a bridge between the opposing alliance-groups, or to mitigate a war, or to facilitate an attempt to lure over a state that was in the other *bloc*.

Sometimes the principle of the balance of power is regarded as a precept, about which one can have a certain option – an injunction to behave in one manner rather than another. The 18th century regarded it rather as a law which operates wherever there is an international order and a states-system, a law which operates if governments are alive to their long-term interests. It is an injunction only in the sense that states are expected to pursue not merely local advantages but also their own long-term interest. Modern historical writing developed out of the literature on the balance of power. The historians of Göttingen adopted the concept of 'the European states-system', taking it over from the publicists and the practitioners of politics. They, too, regarded the system as a modern one, a Western creation, in spite of the fact that conditions somewhat analogous existed in ancient Greece, and in the monarchies that succeeded the empire of Alexander the Great. One of the Göttingen historians, Heeren, went on praising the system at a time when Napoleon, then at the height of his career, had entirely overturned it. Even at this inauspicious moment Heeren suggested that this international order, this European states-system, might gradually be expanded until it embraced the whole globe. Ranke took over from the Göttingen historians this whole idea of a European states-system, as well as the notion of the balance of power and the recognition of the operation of force in the resulting order of things. Englishmen in the 20th century condemned Ranke as a Prussian for this aspect of his teaching, not realising that the Hanoverian school from which he learned it, had developed it precisely as a result of its connection with England.

I should infer from all this that an international order is not a thing bestowed upon by nature, but is a matter of refined thought, careful contrivance and elaborate artifice. At best it is a precarious thing, and though it seems so abstract it requires the same kind of loyalty, the same constant attention, that people give to their country or to the other private causes which only the international order enables them to follow. I should infer, again, that the world was wrong in 1919 when it decided that hitherto everything had been in a state of nature, with only anarchy in the international field – all thought about foreign relations and international systems having to start at the beginning again. I personally would like to know what would have happened if the kind of international thought which we have been examining had received, not permanent establishment, but simply continued development during the last 150 years.

In reality it became crossed with ideological notions of foreign policy which greatly affected the world from the time of the French Revolution. In the 19th century it is interesting to see how revolutionaries and liberals could design to spread their system through an active foreign policy and through war; and there were many Frenchmen in 1793, 1830 and 1848 who would have been shocked to think that by the end of the 19th century a French republic would be allied with Czarist Russia. It seems that the world has constantly to be weaned out of the notion that wars ought to be war of religion and turned – if they exist at all – simply into wars for the creation of an international order.

I should infer finally that a given educational system is an odd one if it does not teach the young the profound ideas and principles which lie at the basis of the order in which they live.

Notes

1. Polybius, book I, ch. 83, 4.
2. *The Prince*, ch. xxi.
3. Letter to F. Vettori, December 1514 (*Opere* (Firenze, Conti, 1818–21), vol. x, pp. 199–211).
4. *Ricordi politici e civili*, no. lxiii; cf. nos. ccxxxvii and ccxxxviii (*Opere Inedite di Francesco Guicciardini* (Firenze, Barbèra, Bianchi, 1857–58), vol. I, pp. 110–11' 170).
5. *Storia d'Italia*, book I, ch. i.
6. 'Of Empire', first published 1612 in the second edition of *Essays* (*Works of Francis Bacon*, ed. J. Spedding, R. L. Ellis and D. D. Heath, vol. vi, p. 420).
7. *Memoirs*, book V, ch. xviii; see below, p. 149.
8. See a letter from Mazarin to the comte d'Avaux, in 1646: 'Le motif principal du Vénitien en cela est peut-être pour mettre les choses en cet équilibre que la République a si fort en tête' (quoted in G. Zeller, *Les Temps modernes, I. de Christophe Colombe à Cromwell (Histoire des Relations Internationales*, ed. P. Renouvin, vol. ii, Paris, Hachette, 1953), pp. 203–4). In another sphere, cf, the treaty of alliance between France and Denmark, 25 November 1645, art. xii: 'Comme la liberté du Commerce consiste principalement en ce que les choses soient maintenues, dans l'Ocean Occidental, dans la Mer Septentrional, et dans la Mer Baltique, au mesme éstat qu'elles ont esté jusqu'à present, l'un et l'autre Roy travaillera et s'employera pour que cet ancien et salutaire équilibre qui a servi jusqu'à present de fondement à la Paix et à la tranquillité publique, soit conservé par tout sans aucune alteration'

(J. Dumont, *Corps Universel Diplomatique* (Amsterdam, 1726–31), vol. vi, part I, p. 329).

9. 'Sur la nécessité de former des alliances, tant offensives que défensives, contre une puissance étrangère qui aspire manifestement à la monarchie universelle', Supplement a *L'Examen de Conscience sur les Devoirs de la Royauté (Oeuvres*, Paris, Lebel, 1824, vol. xxii, pp. 306–15).
10. 'Of Seditions and Troubles' and 'Of Empire' (*Works*, vol. vi, pp. 410–20).
11. F. von Gentz, *Fragments Upon the Balance of Power in Europe* (London, Peltier 1806), p. 63 note.
12. 'Bien que ce soit un dire commun que quiconque à la force a d'ordinaire la raison, il est vrai toutefois que deux pussiances inégales, jointes par un traité, la plus grande court risqué d'être plus abandonée que l'autre,' Richelieu, *Testament Politique*, ed. L. Andre (Paris, Laffont, 1947), p. 354. See Rudolf von Albertini, *Das politische Denken in Frankreich zur Zeit Richelieus* (Marburg, Simon, 1951), p. 188.
13. *Politische Corespondenz Friedrich's des Grossen*, Erganzungsband, *Die Politischen Testamente Friedrich's des Grossen* (Berlin, Hobbing, 1920), p. 48. '*Voilà un équilibre de puissances qui s'établit et qui constitue une égalité de force entre l'agresseur et l'attaqué .... cette politique, une fois établie en Europe, empêche les grandes conquêtes et rend les guerres infructueuses, à moins qu'elles ne soient conduits avec une grande superiorité et une fortune invariable.*' Cf. p. 49 which declares the '*coup d'éclat*', the surprise attack on Silesia, unrepeatable. Frederick's attitude to the problem of peace and war in his Political Testament of 1752 is discussed in E. Bosbach, *Die 'Rêveries Politiques' in Friedrichs des Grossen Politischem Testament von 1752* (Köln Graz, Böhlau, 1960), pp. 64–7.
14. *Annual Register* for 1760, pp. 2–3. Cf. ibid., for 1772, pp. 2–3.
15. *The Decline and Fall of the Roman Empire*, 'General Observations on the Fall of the Roman Empire in the West', following ch. xxxviii.
16. *Fragments upon the Balance of Power in Europe*, ch. I, *ad init.*

Conclusion

A premise of this collection, set out in our Introduction, was that Butterfield's international thought has been rather left in the margins by the heirs to the intellectual project of which he was a co-founder. It is worth pulling them together in a single volume, we argued, because what he has to say is of exceptional interest to a new generation of scholars for whom the problems that concerned him have gained a new importance. In most respects, we hope that the selections presented here speak for themselves to our success in this regard. It is, however, a selection and those who know their Butterfield will undoubtedly be disappointed with our choices. Where, for example, are the essays on *Western Civilization* and *Just War*,[1] and why have we not included an example of his historiographical detective work such as 'The Reconstruction of an Historical Episode: The History of the Enquiry Into the Origins of the Seven Years War'?[2] In our defence, we make the usual excuses of editors about constraints of space, the tyrannies of the publisher in this regard, and the difficulty of including certain pieces because of their length and unsuitability for editing. We can only add that we too are disappointed by what we have left out. Both of us have had to be persuaded to make sacrifices in this regard.

As to the collection speaking for itself, the historians' confidence about good work in this regard has given way, as it has on so many other fronts, to the social scientists' need to make points explicit, spell out the assumptions on which they rest and, in short, belabour insights, sometimes to the point at which the spark of wisdom seems beaten out of them. We have tried to avoid the latter and, to this end, we will not rehearse the key points of Butterfield's contribution to international thought as these were set out in the Introduction. Rather, we will consider several problems to which his arguments constantly lead – problems which he never satisfactorily addressed but which, in our view, press upon the world now even more than at his time of writing.

The first concerns the balance of power as a way of thinking about and practising international politics. What, if anything, does an account in these terms have to tell us about international relations at a time when the notion of political communities living separately in a world in which power is sufficiently distributed, is under sustained assault from claims about rising interdependence, emerging world orders, and projected hegemonies?

The second concerns diplomacy. How are we to obtain good diplomats and, perhaps, more importantly, how are good diplomats to become influential in the formulation of foreign policies and the conduct of international politics and international relations?

Much of the value of historical study, according to Butterfield, resides in discovering how people from other times and places framed and worried about the things over which we spend much of our time worrying. From such discoveries, we may obtain knowledge, understanding and not a little comfort. Butterfield, himself, is rapidly becoming history, and it is in this sense that his contribution to contemporary concerns is best understood. Indeed, those familiar with Butterfield, will know he was no fan of contemporary fashions in the way these are generally understood. Too restrictive a time frame tends to produce a sectionalised view of the processes underlying historical change – ignores the complexities of the human condition and hence oversimplifies the patterns of international behaviour. Here is why Butterfield opposed excessive preoccupation with contemporary history for this fostered parochialism – caused students 'to look themselves more securely in the deceptions and illusions of their age'. He also believed that the incandescent friction between direct utility or application and historical sensitivity must be recognised. If, to his other qualifications, as Butterfield wrote in his Inaugural lecture, a statesman adds training in history, what he gains is rarely the kind of knowledge which, by a system of transfer can safely be applied to the problems of the present day. It is more likely to be something which, merging with the rest of his wisdom, will be discernible only as reflected light, an intuitive sense of current limitations and a sensitivity to the processes of change over time. Engaging contemporary problems as if they had no history or as though, if solved, they would take us beyond history involves the sins of pride and self-righteousness, putting people on the road to unhappiness and much worse. For Butterfield, the important arguments were eternal, only their iterations shifted. Thus it is that questions about interdependence, world order, hegemony and good diplomacy would be, at heart, for him, nothing more than variants of an old problem. How is wisdom to be recognised as such and, once recognised, how do people become encouraged to act upon it? Here, characteristically the point of departure, according to Butterfield again is history – the historical knowledge of how different conceptions of international politics and concomitant styles of diplomacy, evolved across the centuries into their present configuration. Ultimately he believed the *Ancien Régime*, despite its excesses and however removed in time, had devised vital concepts for managing and preserving equilibrium among chronically competitive states: a model of institutional cooperation and framework of thought and action conducive to rational communication, moderation and hence the stabilisation of interstate affairs. In this context, diplomacy emerged as a moderating institution conducive to rational communication, conciliation and the stabilisation of relations among European powers. Within this scenario, international policy and the ordering of state objectives remained flexible, subject to modification in response to changing conditions and needs. The reservoir of diplomatic techniques tended to be extensive, the discretionary powers of diplomats substantial, and diplomacy, employed to attain long range results, enjoyed a vital and pre-eminent role.

However, following the revolutionary wars and throughout the 19th century, the texture of international politics experienced striking change. Ideology and nationalism arose as divisive forces in national life; states became unified social mechanisms, their citizens exerting increasing influence on political decisions, the dynastic interests of monarchs were increasingly fused with national goals and interests, prompting them to pursue their policies to excess. All this seriously disarranged the tenor of traditional diplomacy, undermined the consensus necessary for effective collaboration and limited the opportunities for constructive diplomatic dialogue. Diplomacy came to concentrate on consolidating or extending existing arrangements – not, as formerly, on coordinating and moderating interstate policies in the interests of the European community as a whole. Consequently, as the influence and negotiating scope of diplomats declined, so did their ability to pre-empt the massive military escalation which eventually erupted into the First World War.

The continuing decline of diplomacy as an instrument of international accommodation after 1918, was part of Europe's wider failure to replicate the diplomatic system that had collapsed in 1914 and for many of the same reasons. Dissension among the victor nations, the discontent of the defeated powers, the widening ideological rift between east and west, expansion of membership in the diplomatic community of new states lacking common traditions and experiences. All this, circumscribed the bargaining range within which diplomacy could function: no longer able to modify or reformulate national objectives in response to international interests, it became limited to devising ways in which the fixed agendas of nations – whether territorial, political or ideological – might be adjusted to some degree, generally without reference to the international context or balance of power considerations. The result was a serious weakening of negotiation and a corresponding increase during the interwar years in the primacy of military measures and agendas at the expense of international consensus and order – a tendency still with us today. Yet, according to Butterfield, the earlier precepts that had once fostered systemic stability could be resurrected and reformulated to meet contemporary needs. This meant renewed focus on the ethical and normative dimensions which have held together the society of states and continue to hold it together.

International relations today and the balance of power

Anyone who takes the balance of power seriously, either as an account of the gravitational forces which regulate the relations of independent states or as a way of organizing those relations historically developed for peoples' interested in preserving their independence, must be struck by the extent to which virtually no one agrees with them anymore.[3] Despite the *prima facie* grounds provided by the United States' current ascendancy, together with the fear (and jealousy) to which this gives rise in others, it often seems as though the only people who continue to think in these terms are Americans themselves: some who are on the lookout for challenges to the US position; and others who still see the US committing its 'weight' to maintaining balances around the world. As for the rest, they regard the balance of power as empirically inaccurate and conceptually imprecise or as based on metaphors that do not help us address important problems. Indeed, far from helping, balance of power thinking is widely regarded as making things worse. It wastes resources in a vain attempt to maintain some conception of the balance and is willing to sacrifice peace to maintain both independent states and the order that sustains them. In a world where security threats are generally said to come from extremists expressing or exploiting the anger of those who remain poor and oppressed in the midst of great wealth and the potential for great freedom, and in which the only real arguments are said to be about how best to deliver that wealth and freedom to all in a manner which is sustainable and fair, why would anyone want to reconstitute the great orchestration of the balance of power?

One answer, of course, is to say that the balance of power has never gone away. Just because it is depreciated, like power itself, does not mean it is not there.[4] The claim takes two forms: a 'scientific' and a 'historical' version. Building on the Newtonian metaphors noted by Butterfield, the former parlays glimpses of, for example, Sino-Russian co-operation or French ministers expressing concerns about American 'arrogance' into evidence of great shifts in the tectonic plates of the geopolitical system. In this view, the balance of power appears to have been in abeyance because the world has been in the process of shifting from one balance of power – the Cold War system of the superpowers – to a new one countering the United States.

The scientific version sees the balance of power operating with a high degree of automaticity below the level of psychological consciousness. In contrast, the historical version sees it operating below the surface of what might be best termed popular political consciousness. Some people have never liked the notion, and it has always been contested, the more so in political systems which are legitimised on democratic and popular grounds. Behind the scenes, in this view, those with primary responsibility for national security have continued to think and guide their states in balance of power terms. China's rising and America's attempts to counter it, therefore, are both manifestations of conscious statecraft obscured by a primarily economic discourse of cooperation and competition.

Both saving hypotheses are beset with problems. The evidence used to support them can be read in many ways, and there exists a great deal of evidence that is simply at odds with either view. We live in a world where, as Butterfield might have expressed it, we see some local signs which remind us of the balance of power but which, on the whole, does not operate in such terms. Rightly or wrongly, no one seriously believes that Sino-Russian strategic cooperation agreements or Franco-German collaboration in the UN Security Council, for example, constitute attempts to balance the United States at any more than a rhetorical level. The dominant metaphors of today's international relations do not involve the idea of performatively constituting a great dance valued both for itself and the independent lives made possible by it. Instead, they revolve around the idea of solving problems by collaborative building and development. Rather than seeking out the balance of power as some sort of hidden hand in all this, therefore, critics argue, we should seek to grasp the other end of the stick and attempt to understand its absence. To the extent that an international system results from no more and no less than the way in which individual people come to think about their circumstances, we should be asking why people, especially influential people, have ceased thinking and, thus, acting, in balance of power terms.

Three possibilities present themselves. First, as suggested above, it may be that diplomacy has already failed in terms of Fénelon's supreme purpose, preempting the emergence of a preponderant power. The United States may have already achieved on a world scale the position to which Spain and France aspired in 16th and 17th Century Europe, and the relations of independent states may be in the process of transmuting into the relations of hegemony and even universal empire. Secondly, the American unipolar moment, such as it was, may have already passed, and, far from being in decline, balance of power thinking may be poised for takeoff as the limits to US power become clearer and the consequences of challenging it less daunting. That both these propositions can be made about the same world is, in itself, some sort of judgment on their validity and utility. It is a comfort to be reminded that at any given point in history, contested and contradictory interpretation of what is going on in international relations are the norm, and agreement on which was 'right' emerges only with the passage of time and with whatever benefits hindsight may confer. It seems reasonable to say, however, that present times are unusually unsettled in these terms.

The third possibility, therefore, is that a fundamental change in the character of international relations, and one that does not arise merely from changes in the distribution of power between states, has occurred or is occurring. People's thoughts and actions may not yet be able to give a definitive or settled shape to the world resulting from these changes. However, there exists *prima facie* evidence for claiming that the constraints that press, and the incentives that encourage, people to think in balance of power terms are gone or, at least, greatly reduced. What are these constraints and incentives? The constraint is that people generally understood themselves to be living in independent political communities. The incentive is that they generally understood living this way to be good. From these emerged what may be best regarded as a pragmatic, moral principle that others ought to be allowed to live independently too. We say pragmatic, because the strategy of securing independence by conquering and absorbing all threats to it remained a temptation for some. For the weak and the lazy who could not entertain this option, and the good and the wise who would not entertain it, being ready to collaborate with others against threats to the independence of all offered an attractive way of trying to safeguard independence. Joining with others in an attempt to balance such threats was less costly, painful and risky than succumbing to the imperial temptation and avoided the moral hazards of responding to aspirant hegemons with pacifist or utopian strategies. Such readiness on the part of a sufficient number of their fellows would also provide a powerful deterrent to those states tempted to achieve absolute security by conquest. While everyone might argue about the particular configuration of these independent political communities at any given time (who should be independent, for example, and on what terms?), nearly everyone could agree that the world consisted of independent political communities and that keeping it so was good.

What has changed? Certainly not the idea that people live in independent, separate political communities. Most people, most of the time, think and talk as if this is the case. The changes we see are more easily described in terms of changes in the particular configuration of these independent communities – some merging and some breaking apart – rather than changes in the principle that we live in independent communities. It may be objected that, while we may think we live independently and present ourselves thus, in fact, this is no longer the case. Economic issues, for example, are often presented as demonstrating the 'myth of independence'. How can one be independent if the directions of trade and investment flows, interest rates and exchange rates are effectively beyond one's control? The problem with this sort of question is that it raises the bar on what it means to be independent. It does so by transforming constraints upon the exercise of independence that have always been more or less present into elements of a definition of independence. Certainly, we may speak of many new issues, or issues which are currently enjoying a new salience, in people's understanding of how the world works which fall outside earlier and established conceptions of the proper business of international relations. Mapping these may allow us to talk about the ease or difficulty with which Belgium, for example, exercised different consequences of its independence in 1839, 1914 and1939 but it does not enable us to say anything about whether or not Belgium *was* independent (in the way that looking at Belgium in 1838 and 1940, for example, would).

Nor does it seem that the idea of living in independent political communities being good and natural has changed either. The struggles of the Palestinians and many other nationalist movements bear testimony to its enduring value, even if people have become more sober about what independence can currently achieve for them. The efforts of the European Union to develop its potential as an agent by creating a common European identity above the national identities of its members, illustrate how even profound change remains framed in the national-state model. Indeed, it remains very hard to imagine alternatives to a plural system of independent states that people would regard as achievable and desirable. Establishing the state-solution to order on a global scale is certainly conceivable, but hardly anyone says they want it. Associations of less-than sovereign states occur within the sovereign framework of a federal or confederated state like Germany or Canada, and when, like the former republics of the USSR and Yugoslavia, they get into difficulties, it is assumed that the political options are either a successful re-establishment of sovereignty at the federal level or the emergence of a sub-system of new sovereign states.

If people still want to live independently, do they still accept the moral corollary of being willing to let other people live independently too? This is a much harder question to answer. It may be plausibly claimed that the willingness to let others live independently is far stronger today than it was at times when the operations of the balance of power were far more in evidence. After all, during the latter's heyday, large numbers of people either had been, or were in the process of being, brought under the rule of a handful of empires run by European great powers. Interest, vice and moral reasoning combined to allow the great powers to persuade themselves that these peoples were either not ready to participate fully in political and international life or might only do so once absorbed by one of their number. Today, in contrast, the independence of peoples in states whose sovereignty and boundaries are inviolable, is a principle enshrined in the preambles to most international agreements, and nearly everyone agrees it is right that this be so. We may point to departures from this in practice. Kurds today have as much to say about the value of the principle of national self-determination as Poles did in the past about the principle of preserving the liberties of states. The violation of a principle, however, while indicative of problems with it, also confirms that it exists. Poles eventually benefited from the principle and Kurds may do so yet.

The real threat to the idea of letting others live independently, does not come from the failure of those who espouse it always to live up to that ideal. It comes from another source, the assumption that, if left to themselves, most people will want to live the same way. We can see robust versions of this in those who believe and act as though real freedom will result in everyone living in, for example, a particular conception of market democracies or a particular conception of the Umma of the Faithful. It can also be seen to exist in the populist hunch which we all share to some degree that people are broadly the same the world over or, more subjectively, that people are like us or me. The assumption that truly free or independent people will be the same undermines, not the value we place on other people's independence so much as, our ability to act in such a way as to secure this independence. In its robust or assertive form, it poses a threat to the independence of others and, thus, of all, for the absence of self-restraint can easily translate into an imperial or hegemonic project. In its populist form it inhibits the ability of others to respond to such threats. It is hard to imagine people who are like us, no matter how rich and powerful, posing a threat, while we easily worry about people who are different, no matter how poor and weak they may be. Thus, while countries and other international actors race to collaborate with one another for a host of technical, economic and political reasons, the one thing on which they will not currently collaborate is to counter the preponderance of one of their number. International relations, as a consequence take the form of arguments about how to solve problems or, if the starting assumptions of the participants are too far apart, how to get rid of those with whom one cannot agree or the beliefs which sustain them. There seems to be very little in between.

Clearly, the condition of independent political communities persists, as does the general belief that this is the right way to live. The moral corollary that others must be allowed to live independently also survives, albeit modified by strong and weak forms of the assumption that we are all the same. What has been weakened, however, are the ideas of self-restraint and collaboration that sets power against power independent of the content to which such power might give expression. In Butterfield's terms, we have lost both the wisdom and the virtue that made the balance of power possible, even though we continue to live in conditions that are best constituted as a balance of power. If this is so then how is this wisdom to be recovered and this virtue to be restored?

The restoration of diplomats and diplomacy

People ought to be taught them. This is the answer touched upon, but never developed, throughout Butterfield's writings on international relations. It was not developed for a number of reasons. For all his fascination with the contribution which technique makes to diplomatic practice, Butterfield belonged to a generation and class of academics which, insofar as they thought about educational technique at all, concerned themselves with the content of the courses from which intelligent men and women might draw the truth for themselves. He also probably saw the truths manifested in the balance of power as the sort at which one arrives, in part, by inward reflection and self-understanding rather than by outward observation and measurement. To the extent that this was so, then the balance of power and everything sustaining it belonged to the realm of Providence. Men and women might be given the wisdom to recognize its prerequisites, but they might, through willfulness, bad luck or for reasons unknown to us, lose that wisdom. As a historian and as a Christian, Butterfield had plenty of evidence of these human weaknesses, the dark consequences that might proceed from them, and the importance of lives being lived without placing bets on History seeing us through in the end. Nevertheless, his faith, the attendant hope, and his humility in the face of a myriad of facts upon which no human mind could aspire to impose a single order, kept Butterfield open to the possibility of better things appearing from unexpected quarters. This disposition to keep trying, together with the number of occasions on which he made his calls for a return to the study of diplomatic history, strongly suggest that Butterfield had a genuine, if undeveloped, commitment to international education with a practical view in mind.[5]

While we might all agree on more and better international education, however, this begs two questions. What should be taught and to whom if one is to develop an international system best suited to how people are, how they want to live, and their understanding of both? One thing is reasonably clear; it is not by telling the story from Westphalia to Versailles as the fall of Man from diplomatic grace. As noted earlier, Butterfield made it clear that the restoration of that particular paradise was not possible. The golden age of diplomacy was to be appreciated in the way we might appreciate the architecture of the time. He may have left it to others to point out the shortcomings of 18th century diplomacy (stately homes are drafty and cold too), but Butterfield had no doubt that it belonged to a world that was lost. For him, however, this did not entail that one started all over again identifying new agents and processes by which international relations might be conducted. Instead, as his popular essays on international relations during the Cold War illustrate, Butterfield sought to identify what was unchanging and to demonstrate how it operated under different circumstances. Fear and insecurity, for example, were constants even if they were expressed differently and with different consequences in an era when populist regimes equipped with nuclear weapons and universal ideologies confronted one another.

Today, however, the task of finding and agreeing upon these constants has been greatly complicated by the very expansion of international relations teaching and research for which the British Committee on the Theory of International Politics, among others, originally called. There is no shortage of international education and, as a consequence, we can no longer so easily assume, as Butterfield did, a central story of international relations around which a great deal of other phenomena revolve. There are a great many other stories, many of them focusing on how the world is not, or need not be, as it appears, others demonstrating how people, or some people, cannot or will not behave well, while a few even promise (or threaten) to take us out of the realm of international relations altogether. Wight's presentation of international thought as a three-point discussion between realist, rationalist and revolutionary traditions of thought suggests this is not in itself a new state of affairs.[6] What is new is that more and more people make sense of the world and act in it on the basis of stories rooted in these different traditions. Indeed, they often participate on the basis of scripts that have not historically been their own. Thus, soldiers become peacekeepers and community-builders, humanitarians worry about the interests and status of the organisations for which they act, and business people become guardians of the international public commons and advocates for international public interests.

The consequences of this development are no more evident than in the contemporary education and training of professional diplomats and those in search of international careers. In most diplomatic academies and foreign service schools, the balance in content has shifted from diplomatic theory and practice to other subjects such as economics, psychology, plus the information and organisation sciences. Political science and international relations also make their appearance although both, particularly the theories generated by them, appear the least valued of all, by teachers and students alike.[7] In many respects, this shift reflects broader changes in views about the content and skills to be transmitted by a good liberal education. It has also been abetted by the unwillingness of traditionalists to let go of a narrow conception of diplomacy and their reluctance to accept that, while good diplomats are always a combination of talent and experience, teaching and training can play their parts in accelerating the development of even the most talented. Most of these changes are to be welcomed, therefore, but what has been diminished as a result is the sense of diplomacy as a distinctive practice undertaken by its own professionals. Even skills traditionally associated with diplomacy are fostered and developed by the application of models and concepts from other walks of life, notably the worlds of business management and negotiation. The point is perhaps best made by noting that, while experts in business negotiations, organisational psychology, inter-cultural communication, education, and management sciences routinely participate in the education and training of diplomats, experts in diplomacy are rarely deployed in the same capacity by schools of business, management, marketing or even international public policy. While everyone has a great deal to tell the diplomats, it would seem that no one, including the diplomats, believes we have much to learn from them.

The principal reason for this is that diplomacy is now widely viewed as a secondary or derivative profession, consisting of nothing more than doing usual things under unusual circumstances.[8] This is certainly how it appears to people working from the 'inside' perspective with which most of us function as citizens of national states. To see diplomacy only in these terms, however, involves forgetting that part of its history in which its distinctive international dimension was fully recognised and made explicit for the first time. In many respects, far more important than the great diplomatic dance of 18th century Europe was the emergence of a diplomatic consciousness in the communities of diplomats made possible by the establishment of resident embassies in northern Italy and Western Europe in the previous two centuries. Working together, they generated a distinctive view, first of their collective predicament as strangers at court, secondly of the particular conditions which made their own work possible and easier, and thirdly of the general conditions which would make life in a system of sovereigns more tolerable. Their answer, of course, was the balance of power as presented by Butterfield.

However, it is not their answer which is of interest here so much as the processes by which it was generated, a sustained reflection by diplomats on their own circumstances and roles as the primary bearers of international relations. If we want people to understand international relations, as opposed to the myriad of partial issues, causes and interests which they are about, a good starting point for international education will still be the distinctive aspects of the predicaments in which diplomats find themselves on our behalf today. We might go further and suggest that specifically diplomatic education be directed at encouraging diplomats to talk to one another with a view to reconstituting themselves as a profession with its own priorities and its own sense of the requirements of a well-ordered international system.

Quite what this would result in is very hard to say, not least because the participants would be a much larger and heterogeneous crowd than those who contributed to the reflections of Commynes, Wicquefort, Callières, Pecquet and other contributors to the canon of classical diplomacy.[9] Indeed, one of their first tasks would be to provide new answers to the old question '...who are the diplomats?' The right of representation was historically defined in terms of the legal standing of those to be represented. Could they claim sovereignty and would others recognize it? Claimants had to be public authorities, but power always played a part in sustaining such claims, for the strongest tended to be states and they insisted that the principle of living in states be applied to everybody. Insofar as the right to representation depends on power, then it seems likely that the difficult question of granting this right to associations and enterprises will have to be engaged in the near future. It remains very hard, however, to imagine the results of such reflection. We see non-state representation subsumed in or appended to state delegations at international conferences. We see non-state representation fulfilling traditional diplomatic functions in troubled areas where states have failed and retreated. And we have the historical precedent of merchant trading companies exhibiting state-like features and competencies until they were stripped of them by states. Daunting though the challenge of creating a new international political dispensation might be, however, it seems superior to the likely alternatives suggested by metaphors of bifurcated worlds in which the severed heads of states continue to play their games while international life increasingly goes on above, below and around them.[10]

This speculation takes us beyond the terms in which Butterfield thought about international relations. He saw the states system as possessing, and needing, an almost infinite capacity for expansion in its ability to take account of the perspectives and interests of new members.[11] And so long as a states system survived, then so too would the utility of constituting it in terms of the balance of power developed from the reflections of diplomats. It is very difficult, although not impossible, to imagine a similar set of reflections by diplomats under current conditions leading to their advocacy in this direction. The value assigned to diplomatic thought about international relations by Butterfield persists, however. Insofar as human affairs consist of nothing more than individual personalities freely acting in the world and reflecting on their places within it, then we should be interested in the experience and reflection of those individuals who find themselves between the groups in which we like to live and responsible for the conduct of their relations. The historical evidence, Butterfield argued, demonstrated that such people – diplomats – were better placed than the rest of us to rise above the necessarily partial passions, interests and understandings of what is right which drive us into arguments with one another. His Christian faith in the goodness of people led him to believe that diplomats could put this superior position to good use by advocating the kind of conduct – based on charity, sympathy and restraint – which could moderate disputes and keep them contained. We may not share the faith on which Butterfield's confidence in diplomats and diplomatic thought was based, but it is perfectly possible to base his advocacy on another virtue much easier to share, that of hope. We may hope that people try to be good and advocate goodness in the circumstances in which they find themselves and, thus, that diplomats try to be good and advocate goodness in international conduct. Indeed, if we do not hope in this sense, it is not clear why international thought and education should engage us at all.

Notes

1. File 349 BC 'Western Civilization' is a series of drafts which rehearse Butterfield's views on the origins of history, the Christian emphasis on the individual personality, a notion of historical progress achieved in terms of the growth of 'mental transpositions', the capacity to put one's self in other shoes, and also the impossibility of planting the values and habits of mind of Western Civilization in other parts of the world. They must emerge, if at all, at their own pace. File 275 BC 'Just War' is a series of drafts using the Prussian occupation of Silesia to illustrate a number of themes, for example, the illegitimacy of seeking to assert

a right in a manner which puts the whole system of rights at risk; and the illegitimacy of provoking such an act so that one can wage a war against the very existence of the perpetrator. Butterfield also rehearses his distinction between 'wars for righteousness' and just wars.

2. 'The Reconstruction Of An Historical Episode: The History Of The Enquiry Into The Origins Of The Seven Years War' in *Man On His Past: The Study of the History of Historical Scholarship, Cambridge* 1969 (first published 1955), pp. 142–70.
3. For a recent volume addressing this theme see T.V. Paul, James J. Wirtz and Michel Fortmann (eds), *Balance of Power: Theory and Practice in the 21st Century* (Stanford, 2004).
4. H. J. Morgenthau speaks of the dangerous and dishonest tendency, especially in democracies, to speak and act as if power is not a factor in international relations and politics in general. See *Politics Among Nations* (New York, 1948). Butterfield's own writings on diplomacy also stressed the play of power beneath the surface of rational discourse in negotiations.
5. See, for example, 'The Tragic Element in Modern International Conflict' in *History and Human Relations*, 1951, p. 26; *Man On His Past*, Cambridge, 1969, p. xiv; 'The New Diplomacy and the Historical Diplomacy' *in Diplomatic Investigations* (London, 1966), p. 184,
6. Martin Wight, *International Theory: The Three Traditions* (Brian Porter and Gabriele Wight eds) (Leicester, 1991).
7. We have only informal evidence for this claim from involvement with programmes offering diplomatic training and education.
8. See, for example, the official and semi-official studies which appeared in the 1990s worrying about the continued relevance and effectiveness of diplomacy and diplomats: *Re-inventing Diplomacy in the Information Age*, Center for International and Strategic Studies (Washington, 1998) and *Diplomacy: Profession in Peril?* (Wilton Park, 1997).
9. For selections with comment from these and other writers see G.R. Berridge (ed.), *Diplomatic Classics: Selected Texts from Commynes to Vattel* (Basingstoke, 2004).
10. See, for example, the works of James N. Rosenau and especially *Turbulence in World Politics* (Princeton, 1990).
11. See, for example, 'Christianity and Global Revolution' in *International Conflict in the Twentieth Century: A Christian View* (New York, 1960), pp. 101–20; 'Morality and Order' in Porter, pp. 345–50.

Index